Topics in Applied Physics Volume 3

Topics in Applied Physics   Founded by Helmut K. V. Lotsch

Volume 1   **Dye Lasers**   Editor: F. P. Schäfer

Volume 2   **Laser Spectroscopy** of Atoms and Molecules
Editor: H. Walther

Volume 3   **Numerical and Asymptotic Techniques in Electromagnetics**
Editor: R. Mittra

Volume 4   **Interactions on Metal Surfaces**   Editor: R. Gomer

Volume 5   **Mössbauer Spectroscopy**   Editor: U. Gonser

Volume 6   **Picture Processing** and **Digital Filtering**
Editor: T. S. Huang

Volume 7   **Integrated Optics**   Editor: T. Tamir

Volume 8   **Light Scattering in Solids**   Editor: M. Cardona

Numerical and Asymptotic Techniques in Electromagnetics

Edited by R. Mittra

With Contributions by
F. J. Deadrick R. F. Harrington W. A. Imbriale
C. A. Klein R. G. Kouyoumjian E. K. Miller
R. Mittra W. V. T. Rusch

With 112 Figures

Springer-Verlag Berlin Heidelberg New York 1975

Professor RAJ MITTRA

University of Illinois at Urbana-Champaign, College of Engineering
Department of Electrical Engineering, Electromagnetics Laboratory
Urbana, IL 61801, USA

ISBN 3-540-07072-9 Springer-Verlag Berlin Heidelberg New York
ISBN 0-387-07072-9 Springer-Verlag New York Heidelberg Berlin

Library of Congress Cataloging in Publication Data. Mittra, Raj. Numerical and asymptotic techniques in electromagnetics. (Topics in applied physics; v. 3.) Includes bibliographical references and index. 1. Antennas (Electronics). 2. Electromagnetic waves. 3. Electric engineering – Mathematics. I. Deadrick, F. J. II. Title. TK7871.6.M59. 621.38'0283. 74-30146.

Monophoto typesetting, offset printing and bookbinding: Brühlsche Universitätsdruckerei, Gießen

Preface

During the past two decades a large number of books have been written on the general subject of electromagnetics. Most of these publications have dealt with the classical approaches to solving electromagnetic boundary value problems. There are only a few notable exceptions to these, for instance, a monograph by HARRINGTON on the application of the moment method for the solution of field problems and a text by MITTRA on computer techniques for solving electromagnetic scattering and radiation problems. Since the appearance of these two books much progress has been made in the areas of computerated electromagnetics and the application of ray-optical techniques in the low-frequency and high-frequency regions, respectively. This book attempts to present a comprehensive description of some of these important recent developments and to illustrate the application of these techniques to a variety of problems of practical interest, e.g., design of dipole arrays and reflector antennas. Almost all of the material appearing in the book is relatively new and has not appeared elsewhere in any other publication except in the form of journal articles. It is hoped that the book will serve to fill the gap that exists in the current literature on numerical and asymptotic techniques in electromagnetics and will be found useful both as a convenient reference and as a practical tool for investigating electromagnetic radiation scattering problems. All of the contributing authors are well known throughout the world for their many contributions on moment methods, numerical aspects of computerated solution, ray-optical techniques and other topics covered in the book and they have made every attempt to present the material in the text in a coherent and easily readable form. It is their fervent hope that the reader will not only find the book useful as a tool for modern electromagnetic analysis and design, but will also enjoy the clarity of presentation of the advanced material discussed in the book.

Finally, the editor (R. M.) would like to thank all of the authors for their excellent cooperation and the publishers for their extremely efficient production schedule.

Urbana, Illinois, January 1975 R. MITTRA

Contents

Contributors

DEADRICK, FREDERICK J.

Lawrence Livermore Laboratory, P.O. Box 808, Livermore, CA 94550, USA

HARRINGTON, ROGER F.

Syracuse University, Electrical Engineering Department, Syracuse, NY 13210, USA

IMBRIALE, WILLIAM A.

TRW Systems Group, One Space Park, Redondo Beach, CA 90278, USA

KLEIN, CHARLES A.

University of Illinois at Urbana-Champaign, College of Engineering, Department of Electrical Engineering, Electromagnetics Laboratory, Urbana, IL 61801, USA

KOUYOUMJIAN, ROBERT G.

The Ohio State University, Department of Electrical Engineering, 2015 Neil Avenue, Columbus, OH 43210, USA

MILLER, EDMUND K.

Lawrence Livermore Laboratory, P.O. Box 808, Livermore, CA 94550, USA

MITTRA, RAJ

University of Illinois at Urbana-Champaign, College of Engineering, Department of Electrical Engineering, Electromagnetics Laboratory, Urbana, IL 61801, USA

RUSCH, WILLARD V. T.

University of Southern California, Department of Electrical Engineering, Los Angeles, CA 90007, USA

1. Introduction

R. Mittra

In the recent past, much of the effort in electromagnetics research has been directed toward two broad areas, viz., numerical and asymptotic techniques for solving boundary value problems arising in antennas, scattering, propagation, an so on. Not surprisingly, both of these areas are relatively new in the sense that the significant advances in these areas have been made only within the last fifteen years or less. The advent of the computer triggered a prolific surge of interest in the numerical solution of many practical electromagnetic analysis and synthesis problems that were previously regarded as too complex to be analytically tractable using the available theoretical methods. Ever since the publication of the now classic paper "Matrix Methods for Field Problems" by Harrington [1.1] there has been an almost exponential growth of the number of publications in the area of computerated solution of electromagnetic problems. The genesis of modern asymptotic electromagnetics is the pioneering work by Keller [1.2] on the Geometrical Theory of Diffraction, or GTD in abbreviated form. Though the growth of research effort on GTD has not been nearly as dramatic as in numerical electromagnetics, the advances have been highly significant nonetheless.

As with any field which has seen a rapid expansion in a relatively short period of time, the results of recent research efforts in electromagnetics can only be located in highly specialized journal publications which may not be easily accessible to the interested user in the form of a convenient collection. In addition, presentation in these papers may be too abstruse for an average reader to receive any benefit. The only exceptions are two relatively recent publications on numerical electromagnetics, viz. a monograph entitled "Field Computation by Moment Methods" by Harrington [1.3] published in 1968 and a text edited by Mittra [1.4] which bears the title "Computer Techniques for Electromagnetics". No such comparable text or monograph currently exists on the asymptotic techniques for electromagnetics, or, more specifically on the GTD methods.

The present book represents an effort to at least partially fill in this gap, by attempting to present the basic tools of numerical and asymptotic techniques available to the user in a coherent and convenient form. Since the basic principles and applications of the moment method may be found in [1.3], and a host of applications of this technique to the problems of wire antennas, and general three-dimensional scattering from solid surfaces has been adequately covered in [1.4], the topics of discussion in the present text have been carefully selected to avoid duplication of the material already published. Instead, this book deals with the important topic of the application of the moment method to antenna arrays in Chapter 2, and the problem of numerically deriving the characteristic modes of antennas and scatterers in Chapter 3. The knowledge of the modes for a given body shape is not only directly useful for solving radiation and scattering-problems from these bodies, but it also provides great insight into the mechanism of such radiation. Hence, the characteristic mode concept is useful for investigating the antenna synthesis problems as well.

Chapters 4 and 5 of this book examine various computational aspects of the moment method. These chapter discuss the problems of numerical modeling, dependence of the numerical solution on the manner in which the integral equation is transformed into a matrix equation, questions of convergence, stability of matrix solution, difficulties at interior resonant frequencies, and so on. Various computational aspects which may affect the validity of a numerical solution for a thin wire structure are considered in Chapter 4 and some additional questions pertaining to numerical analysis are discussed in Chapter 5. Data based on extensive numerical experiments with the method of moments are presented in these chapters and some important conclusions are drawn on the basis of these results.

Although the two currently available texts [1.3, 4] serve as excellent background material for the discussion on numerical aspects of electromagnetics presented in Chapters 2 through 5 of this book, as mentioned earlier no such counterpart is available in the literature on the subject of asymptotic solutions in electromagnetics. Thus, an entire chapter, viz. Chapter 6, has been devoted to developing the background on ray techniques and bringing the reader up-to-date on the latest methods available for constructing ray-optical solutions of radiation and scattering problems. It is believed that this chapter fills an important need in providing the background and reference material, which is essential for following the proliferation of recent journal publications. This chapter also serves as an excellent introduction to Chapter 7 which deals with the topic of reflector antennas and which makes extensive use of ray optical techniques for computing the characteristics of reflector antennas that are typically very large in terms of the illuminating wavelengths.

It is a well recognized fact that the numerical techniques are limited in application to radiation and scattering from bodies that are not electrically large. This is because the precipitous growth of the matrix size and the associated processing time on the computer become prohibitive at smaller wavelengths, that is in the resonance region or above. In contrast, the ray-optical techniques which are high-frequency asymptotic solutions, obviously work best when the body size is large compared to the wavelength. Thus, in providing the analytical and numerical tools for solving the electromagnetic problems by approaching the spectrum from both ends, the book will hopefully serve the needs of a broad range of users who would obviously benefit from the wide array of techniques presented in the text. Also, it is hoped that the availability of a text such as this will spark future research interest in electromagnetics with the result that numerical and asymptotic electromagnetics will be brought even closer together, and perhaps a new arsenal of methodologies that bridge the gap between the low and high frequency techniques will emerge from this happy merger.

References

1.1. R. F. Harrington: Proc. IEEE **55**, 136 (1967).
1.2. J. B. Keller: J. Opt. Soc. Am. **52**, 116 (1962).
1.3. R. F. Harrington: *Field Computation by Moment Methods* (The Macmillan Company, New York, 1968).
1.4. R. Mittra (editor): *Computer Techniques for Electromagnetics* (Pergamon Press, New York, 1973).

2. Applications of the Method of Moments to Thin-Wire Elements and Arrays

W. A. IMBRIALE

With 32 Figures

The use of high-speed digital computers with recently developed numerical methods makes possible the solution of many electromagnetics problems of practical interest which, by classical techniques, would be virtually impossible. A major advantage of numerical techniques is that they may be applied to a body of arbitrary shape and are generally only limited by the wavelength size of the body. This limitation is practical rather than theoretical, in that a set of linear equations describing the system can be generated but this set may be too large to be solved in a convenient manner. A principle objective in solving a large electromagnetics problem is to obtain a minimum set of equations for a given degree of accuracy.

The purpose of this chapter is to present a practical method for solving antennas constructed with arrays of wire elements.

Wire antennas are solved using a moments solution where the method of subsectional basis is applied with both the expansion and testing functions being sinusoidal distributions. Using sinusoidal basis functions the terms of the generalized impedance matrix become equivalent to the mutual impedances between the subsectional elements. These basis functions are extremely useful for the analysis of large arrays of dipoles since the use of one subsegment per dipole is equivalent to the induced EMF method of calculating mutual impedances and gives a physically meaningful results. For an array of N dipoles this allows the minimum matrix size of $N \times N$ to achieve a good first-order approximation to the solution. Of the many applications discussed, this first-order approximation is shown to be an adequate representation of the solution.

Section 2.1 examines the numerical convergence of the moments solution using sinusoidal basis functions. Generally an equivalent radius is used in the evaluation of the self-impedance term to reduce computation time. It is shown that only for very thin subsegments is the correct equivalent radius independent of length and that the use of an incorrect self-term is responsible for the divergence of numerical solutions as the number of subsections in increased.

Section 2.2 discusses the solution of single and multiple Log-Periodic Dipole (LPD) antennas. The analysis is formulated in terms of impedance and admittance matrices for the dipole and transmission line networks. The first order approximation is shown to be equivalent to the original solution proposed by CARRELL [2.1] and is adequate for the practical range of operation of LPD antennas. In addition, the analysis for the detailed characteristics of the antenna over a wider range of operating conditions is provided by the complete moments solution. Also, the effects of feeder booms on the performance of multiple LPD antennas is examined.

Section 2.3 combines the method of moments for wire antennas with the physical optics scattering from reflectors to provide a description of LPD antennas as feeds for parabolic reflectors. Utilizing the numerical solutions, design curves for optimizing the gain of LPD fed reflectors are generated. In addition, a technique for reducing mutual coupling between multiple LPD feeds is presented.

2.1. Method of Moments Applied to Wire Antennas

The method of moments is applied to wire antennas, as discussed in other papers [2.2, 3], but carried to a higher order of approximation which allows treating the case where the length to radius ratio is small. The theory will address the straight wire antenna but the extension to wires of arbitrary shape is straightforward.

In the moments solution the method of subsectional basis is applied with both the expansion and testing functions being sinusoidal distributions [2.4]. This allows not only a simplification of near-field terms but also the far-field expression of the radiated field from each segment, regardless of its length. Using sinusoidal basis functions, the terms of the generalized impedance matrix are the mutual impedances between the subsectional elements and can be computed using the induced EMF method.

In the induced EMF method an equivalent radius is usually used in the evaluation of the self-impedance term to reduce computation time. It is shown, however, that only for very thin subsegments is the correct equivalent radius independent of length. When the radius to length ratio (a/L) is not small, an expansion for the equivalent radius in terms of a a/L is given for the self-impedance term. The use of incorrect self term, obtained by using a constant equivalent radius term, is shown to be responsible for divergence of numerical solutions as the number of subsections is increased. This occurrence is related to the ratio of a/L of the subsections and hence becomes a problem for moderately thick wire antennas even for a reasonably small number of subsegments per wavelength. Examples are given showing the convergence with the correct self terms and the divergence when only a length independent equivalent radius is used.

Sinusoidal basis functions are extremely useful for the analysis of large arrays of dipoles since the use of one subsegment per dipole is equivalent to the induced EMF method of calculating mutual impedances and gives a physically meaningful result. For an array of N dipoles this allows the use of the minimum matrix size of $N \times N$ to achieve a good "first-order" approximation to the solution. If n subsegments are required to characterize the behavior of each dipole in the array then the minimum matrix size required for a solution would be $n \cdot N \times n \cdot N$. For large N and moderate n this matrix size quickly exceeds the capability of most computer systems. Of the many applications discussed in later sections, the use of the "first-order" approximation is shown to be an adequate representation of the solution.

2.1.1. Basic Theory

Figure 2.1 shows a straight section of wire of circular cross section, and defines the coordinate system. The wire extends from $z=0$ to $z=L$ along the z-axis and is of radius a. It is assumed that the radius is small compared to a wavelength but the ratio of a to L need not be small. The only significant component of current on the wire is then the axial

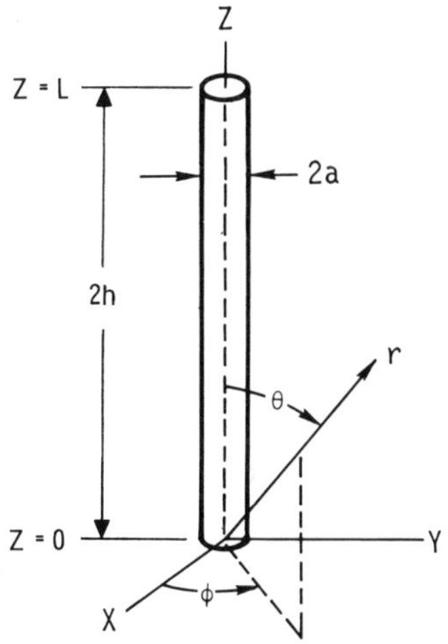

Fig. 2.1. Straight wire and coordinate system

component, which can be expressed in terms of the net current $I(z)$ at any point z along the wire. The current distribution will then be modeled as an infinitely thin sheet of current forming a tube of radius a, with the density of current independent of circumferential position on the tube.

An operator equation for the problem is given by

$$L\{I(z)\} = j(4\pi\omega\varepsilon)^{-1}(d^2/dz^2 + k^2) \oint_c \int_0^L [\exp(-jkR)/R] \cdot I(z')\,dz'\,dc = E_z^i(z)\,, \tag{2.1}$$

where $E_z^i(z)$ is the z component of the impressed electrical field at the wire surface, $I(z')$ is surface current density, $\oint_c dc$ represents the integration around the circumference, and R is the distance from the source point to the field point.

The procedure is basically one for which the wire is divided into subsections, and a generalized impedance matrix $[Z]$ obtained to describe the electromagnetic interactions between subsections. The problem is thus reduced to a matrix one of the form

$$[Z][I] = [V]\,, \tag{2.2}$$

where $[I]$ is related to the current on the subsections, and $[V]$ to the electromagnetic excitation of the subsection.

Matrix inversion is a simple procedure for high-speed digital computers, and hence the problem is considered solved once a well-conditioned matrix $[Z]$ is obtained. Of considerable importance is the ease and speed of evaluating the matrix elements and the realization of a well conditioned matrix $[Z]$.

The solution to be described uses sinusoidal subsectional currents and Galerkin's method [2.5], which is equivalent to the reaction concept [2.6], and the variational method [2.7]. Let the wire be broken up into N segments each of length $2H$ and let $I(z)$ be expanded in a series of sinusoidal functions

$$I(z) \approx \sum_{n=1}^{N-1} I_n S(z - nH)\,, \tag{2.3}$$

where I_n are constants and

$$S(z) = \begin{cases} \sin k(H - |z|), & |z| < H \\ 0, & |z| > H\,. \end{cases} \tag{2.4}$$

Substituting (2.3) into (2.1) and using the linearity of L, one has

$$\sum_{n=1}^{N-1} I_n L\{S(z-nH)\} \approx E_z^i(z)\,. \tag{2.5}$$

Each side of (2.5) is multiplied by $S(z-mH)$, $m=1,2,\ldots,N-1$, and integrated from 0 to L on z. This results in the matrix of (2.2), where the elements of $[I]$ are I_n, those of $[Z]$ are

$$Z_{mn} = \int_0^L S(z-mH)\, L\{S(z-nH)\}\, dz \tag{2.6}$$

and those of $[V]$ are

$$V_m = \int_0^L S(z-mH)\, E_z^i(z)\, dz\,. \tag{2.7}$$

In solving thin wire antennas, the integration around the current tube is normally removed by replacing the integral with the value of the integrand at one point. This then reduces the equation to a single integral and obviates the singularity of the integrand which occurs when the source and field points coincide during the calculation of the self and first adjacent mutual terms. The singularity is, of course, integrable; and by suitably expanding the integrand, special series for these terms can be obtained and the integral performed in closed form. However, many authors have used an "average" value equal to the radius a. This approximation is described as assuming the current to be totally located on the center axis and the distance a is used to represent an average distance from the current filament to the true current surface. One of the purposes of the following is to show that the value used in this approximation is critical to the convergence of the solution as the number of subsegments is increased and the ratio of H to a becomes comparable to unity. In fact, the use of a single value of the equivalent radius will be shown to be incorrect at any time, but less important for a small radius. To show this, first consider the evaluation of the general term Z_{mn} of an infinitely thin current filament. Since $S(z)$ is the same sinusoidal function used in evaluating radiation and impedances via the induced EMF method and $S(z)$ is the z-directed electric field radiated by the subsectional dipole, it can easily be shown that [2.8, 9] Z_{mn} is given by

$$\begin{aligned} Z_{mn} = 30 \int_{H_{m-1}}^{H_{m+1}} & [-j\exp(-jkR_1)/R_1 - j\exp(-jkR_2)/R_2 + 2j\cos kH_n \\ & \cdot \exp(-jkR_0)/R_0] \cdot \sin[k(H_m - |z|)]\, dz\,, \end{aligned} \tag{2.8}$$

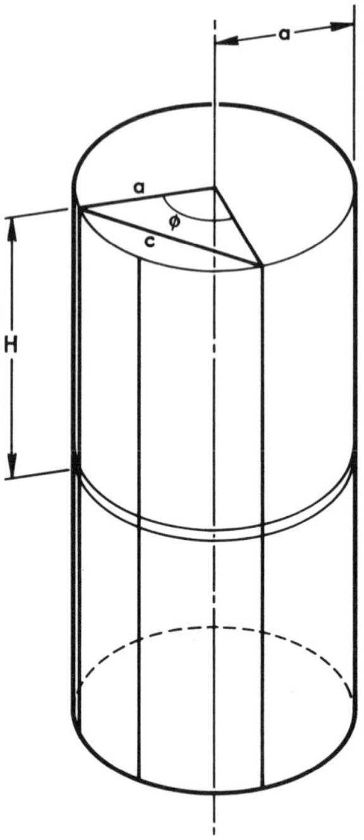

Fig. 2.2. Geometry for self-term calculation

where R_1 and R_2 are the distances from the end points, and R_0 the distance from the center of subsegment H_n to the field point on H_m when integrating over subsegment H_m.

The only problem occurs for the self term and the first adjacent subsegment when the source and field points coincide and hence the impedance calculation for infinitely thin dipoles would yield a value of infinity. It is evident that in computing the value of these terms, the finite diameter of the antennas will have to be considered.

The self impedance of the finite diameter subsegment can be accomplished as follows: Consider the finite segment to be made up of a number of very thin strips of height $2H$ arranged in a circle of radius a as shown in Fig. 2.2. The strips are all assumed to be center fed with a voltage V; hence, the voltage can be written:

$$V = Z_{11} I_1 + Z_{12} I_2 + \cdots + Z_{1n} I_n . \tag{2.9}$$

Since the currents $I_1, I_2, \ldots, I_n$ are all identical and are equal to the total current on the dipole divided by the width of the strip and the impedances

$Z_{11}, Z_{12}, \ldots, Z_{1n}$ are the self and mutual impedances of the thin strips, (2.9) can then be rewritten as a sum over the impedances.

$$V = (2\pi)^{-1} I_{\mathrm{T}} \, d\phi \sum_n Z_{1n} . \tag{2.10}$$

Transferring the sum into an integral using the impedances as a function of $c = \sqrt{2}a\sqrt{1-\cos\phi}$ and also using the definition of self-impedance as being the voltage divided by the current, we have

$$Z_{\mathrm{self}} = V/I_{\mathrm{T}} = (2\pi)^{-1} \int_0^{2\pi} Z(c) \, d\phi , \tag{2.11}$$

where $Z(c)$ is the mutual impedance between two infinitely thin dipoles separated by a distance c.

2.1.2. Self-Impedance Evaluation

It is possible for special cases to come up with a closed form approximation for the self impedance of a subsegment. First, consider the very thin wire case where $a \ll 1$ and $a \ll H$. Writing the impedance $Z = R + jX$ and considering the reactive part first, we can approximate for small c the reactance as follows

$$\begin{aligned} X(c) = -30\{&\sin(2kH)[-\gamma + \ln(H/kc^2) + 2\mathrm{Ci}(2kH) - \mathrm{Ci}(4kH)] \\ &- \cos(2kH)[2\mathrm{Si}(2kH) - \mathrm{Si}(4kH)] - 2\mathrm{Si}(2k\mathrm{H})\} \end{aligned} \tag{2.12}$$

where $\gamma = 0.5772\ldots$ is Euler's constant, and Si and Ci are sine and cosine integrals. Substituting (2.12) into (2.11) and integrating we have

$$X_{\mathrm{self}} = (2\pi)^{-1} \int_0^{2\pi} X(kc) \, d\phi \cong X(ka) \tag{2.13}$$

Equation (2.13) states that for thin dipoles the self-reactance can be obtained by evaluating the mutual reactance between two infinitely thin dipoles at a distance equal to the radius.

If we do the same for R_{self} [2.10] we find that $R_{\mathrm{self}} = R(\sqrt{2}ka)$, which states that the self-resistance can be obtained by evaluating the mutual resistance between two infinitely thin dipoles at a distance equal to $\sqrt{2}a$ and is the same result as obtained in [2.8, p. 361] making use of the induced EMF method for calculating dipole impedances. However, since the resistance is fairly insensitive to radius, approaching

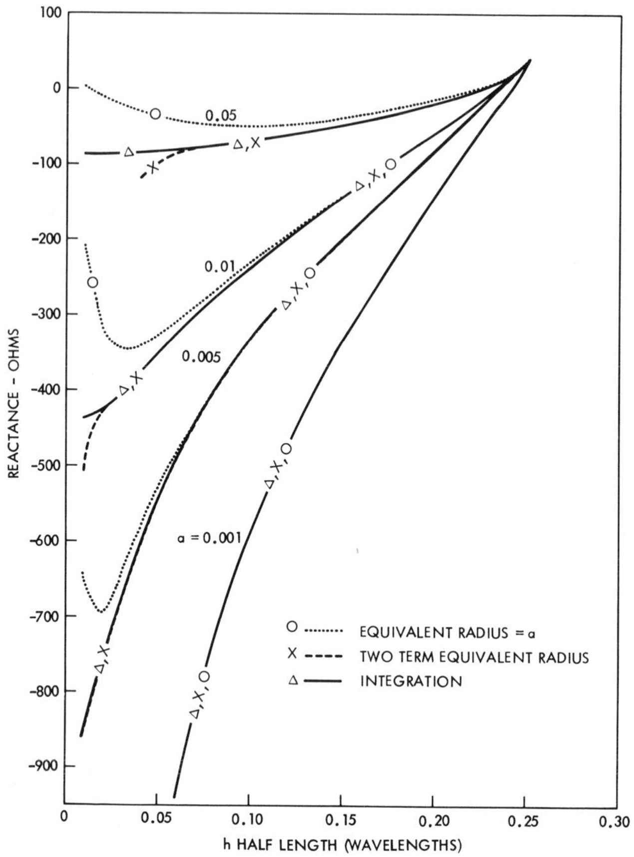

Fig. 2.3. Reactance of a dipole versus half-length comparing the results of integration and two term equivalent radius

a finite limit as $a \to 0$, utilizing the distance a in the calculation of the self-impedance gives excellent results for thin wires.

Continuing on for the case where the radius is still small compared to a wavelength ($a \ll 1$), but comparable to the height H, we can expand the reactance in a series of (c/H) as follows

$$X(c) \simeq A_1 - B_2[\ln(H/c)^2 + 3(c/H) + \cdots], \tag{2.14}$$

where A_1 and B_2 are constants that are not dependent on (c/H). Substituting (2.14) into (2.11) and integrating, we obtain:

$$X_{\text{self}} = A_1 - B_2[\ln(H/a)^2 + (12/\pi)(a/H) + \cdots]. \tag{2.15}$$

Seeking an expansion of the equivalent radius (i.e., a value of $c = a_e$ which will make (2.14) and (2.15) equal) in terms of a/H we find

$$a_e = a(1 - 0.40976\, a/H + \cdots). \tag{2.16}$$

Figure 2.3 is a plot of the reactance of a dipole versus half-length comparing the results of using the first two terms of the equivalent radius series (2.16) to the result obtained when the actual integration (2.11) is performed. For comparison, the corresponding values using a constant equivalent radius are also shown.

2.1.3. Convergence of Moments Solutions

The importance of the utilization of the correct equivalent radius and its effect on the relative convergence of moments solutions is displayed in the following examples.

As a rather extreme case, let us choose a half-wavelength linear dipole with $h/a = 12.5$. When the dipole is divided into 12 segments the H/a per segment is approximately equal to unity. The calculation of the input impedance for various equivalent radii and different numbers of subsegments is shown in Fig. 2.4. Note that for this h/a it was necessary to use the complete integration for the self-term to get a solution that

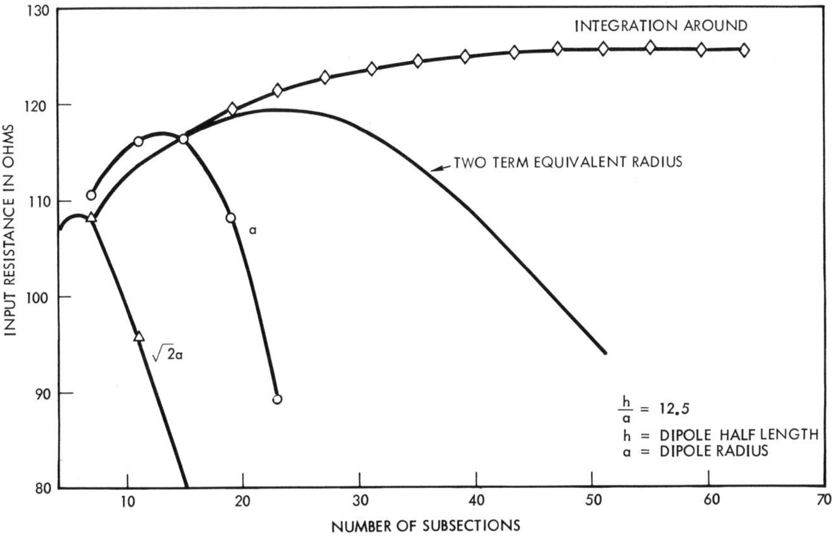

Fig. 2.4. Input resistance of a 0.5 λ dipole with $h/a = 12.5$

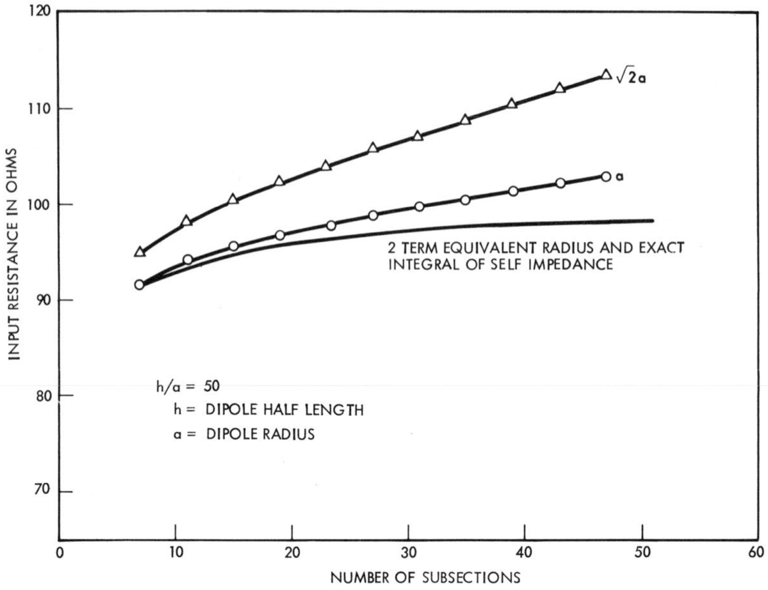

Fig. 2.5. Input resistance of a 0.5 λ dipole with $h/a = 50.0$

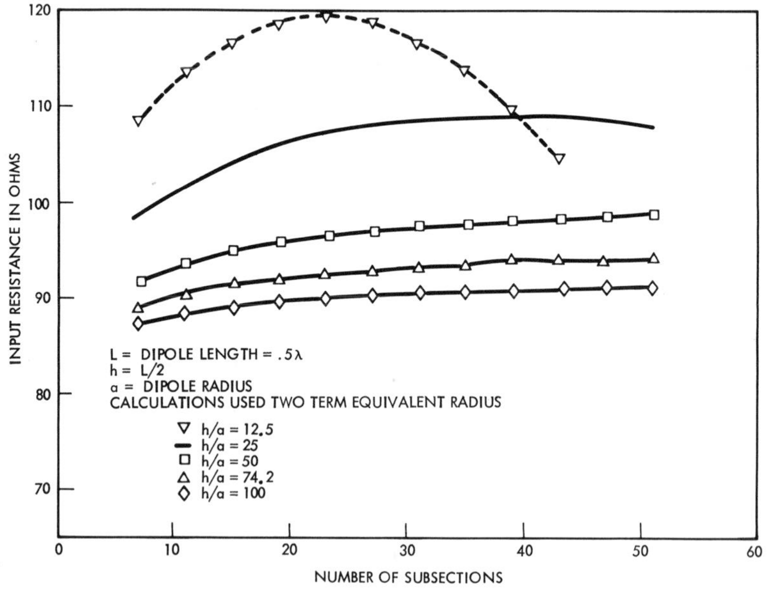

Fig. 2.6. Input resistance of 0.5 λ dipoles versus number of subsections

converged as the number of subsegments increased. Using an equivalent radius of $\sqrt{2}a$, a, and even the two term equivalent radius was not sufficient to yield a convergent solution.

The same type of plot (Fig. 2.5) is made for $h/a = 50$. Here, the two term equivalent radius is sufficient, whereas using a and $\sqrt{2}a$ is not. Plots of the input impedance versus number of subsegments for various h/a ratios are shown in Fig. 2.6 using the two term equivalent radius for the self-term.

2.1.4. Far-Field Evaluation

The radiation pattern of a wire antenna is obtained by superposition of the fields of the many small subsegments with sinusoidal current distributions. Utilizing the general expression for the electric field of a subsegment of any half length H oriented along the z-axis from JORDAN [2.8, p. 319] the far-zone field is given by

$$\begin{aligned} \boldsymbol{E}(\theta, \phi) = j\eta(4\pi r)^{-1} \exp(-jkr) \sum_{n=1}^{N} I_n[\cos(kH\cos\theta) - \cos kH] \\ \cdot \exp(jknH)/\sin\theta\, \hat{\boldsymbol{u}}_\theta\,, \end{aligned} \tag{2.17}$$

where η is the intrinsic impedance of free space and $\hat{\boldsymbol{u}}_\theta$ is a unit vector.

The power gain pattern of the radiation field is

$$g(\theta, \phi) = 4\pi r^2 \eta^{-1} |E(\theta, \phi)|^2 / P_{\text{in}}\,, \tag{2.18}$$

where P_{in} is the power input to the antenna

$$P_{\text{in}} = Re[\tilde{V}][I^*]\,, \tag{2.19}$$

where $[\tilde{V}]$ denotes the transpose of $[V]$, and * denotes conjugation.

2.1.5. Arbitrarily Shaped Wires and Wire Junctions

The procedure for solving arbitrarily shaped wires is similar to that used for straight wire as the wire is divided into subsections, over each of which a sinusoidal current distribution is assumed, and a generalized impedance matrix $[Z]$ obtained to describe interactions between subsections. The junction of two or more straight segments can be thought of as the intersection of two or more half-subsegments superimposed on one another. Thus Kirchoff's current law is not invoked at the junction; it is a consequence of Maxwell's equations.

To complete the discription of arbitrarily shaped wires we need to obtain the mutual impedances between two full subsegments, between a a full subsegment and a half subsegment, and between two half subsegments. The general expression for the mutual impedance is given by

$$Z_{mn} = \int ds \int ds' \, \hat{\boldsymbol{u}}_m \cdot \underset{\sim}{\boldsymbol{\Gamma}} \cdot \hat{\boldsymbol{u}}_n \sin k(H_m - |s|) \sin k(H_n - |s'|) \,, \tag{2.20}$$

where the free-space tensor Green's function is given by

$$\underset{\sim}{\boldsymbol{\Gamma}} = (j\omega\varepsilon_0)^{-1} (k^2 \underset{\sim}{\boldsymbol{I}} - \nabla\nabla') \, G \,, \tag{2.21}$$

$$G = (4\pi)^{-1} \exp(-jk|\boldsymbol{r} \cdot \boldsymbol{r}'|)/|\boldsymbol{r} - \boldsymbol{r}'| \,. \tag{2.22}$$

We shall evaluate the integral (2.20) involving the primed coordinates first. Let the integral be denoted by $\boldsymbol{E}_{mn}$:

$$\boldsymbol{E}_{mn} = \int ds' \, \underset{\sim}{\boldsymbol{\Gamma}} \cdot \hat{\boldsymbol{u}}_n \sin k(H_n - |s'|) \,. \tag{2.23}$$

Substituting (2.21) into (2.23) and after some manipulations, we get

$$\boldsymbol{E}_{mn} = \hat{\boldsymbol{u}}_z E_z + \hat{\boldsymbol{u}}_\varrho E_\varrho \,, \tag{2.24}$$

where

$$E_z^{\pm} = (j\omega\varepsilon_0)^{-1} \int \partial/\partial z' (-G \, \partial f_n^{\pm}/\partial z' + f_n^{\pm} \, \partial G/\partial z') \, dz' \,, \tag{2.25}$$

$$E_\varrho^{\pm} = (j\omega\varepsilon_0)^{-1} \int (\partial^2 G/\partial z \, \partial \varrho) \, f_n^{\pm} \, dz' \,, \tag{2.26}$$

with

$$f_n^{\pm} = \sin k(H_n \mp |s'|) \,. \tag{2.27}$$

The positive and negative signs denote respectively, the upper and lower halfdipoles. Evaluation of (2.25) is straightforward, and evaluation of (2.27) requires a trick given in JORDAN [2.8, pp. 342–369]. Without writing down the gory mathematical detail, the results of the integration are

$$\begin{aligned} E_z^{+} = k(4\pi j\omega\varepsilon_0)^{-1} \{ & \exp(-jkR_1)/R_1 - \exp(-jkR_0)/R_0 \\ & \cdot [\cos kH_n + jz \cdot (1 + (jkR_0)^{-1}) \sin kH_n] \} \,, \end{aligned} \tag{2.28}$$

$$\begin{aligned} E_z^{-} = k(4\pi j\omega\varepsilon_0)^{-1} \{ & \exp(-jkR_2)/R_2 - \exp(-jkR_0)/R_0 \\ & \cdot [\cos kH_n - jz \cdot (1 + (jkR_0)^{-1}) \sin kH_n] \} \,, \end{aligned} \tag{2.29}$$

$$E_\varrho^+ = -k(4\pi j\omega\varepsilon_0 y)^{-1}\{(z-H_n)\exp(-jkR_1)/R_1 - \exp(-jkR_0)/R_0$$
$$Z\cos kH_n + (jz^2R_0^{-1} - k^{-1} + z^2(kR_0^2)^{-1})]\sin kH_n\}, \quad (2.30)$$

$$E_\varrho^- = -k(4\pi j\omega\varepsilon_0 y)^{-1}\{(z+H_n)\exp(-jkR_2)/R_2 - \exp(jkR_0)/R_0$$
$$Z\cos kH_n - (jz^2R_0^{-1} - k^{-1} + z^2(kR_0^2)^{-1})]\sin kH_n\}, \quad (2.31)$$

where R_0, R_1, R_2, y and z are defined in Fig. 2.7 in which the centers of the two subsegments are coplanar with the yz plane. In general, only the following data are given on the two subsegments: the coordinates of the center ($\boldsymbol{r}_i$), the orientation (θ_i, ϕ_i), the half-length (H_i), and whether it is a half or a full subsegment. To express R_0, R_1, R_2, y, and z in terms of the data given, coordinate transformations are needed, the details of which are left to the reader.

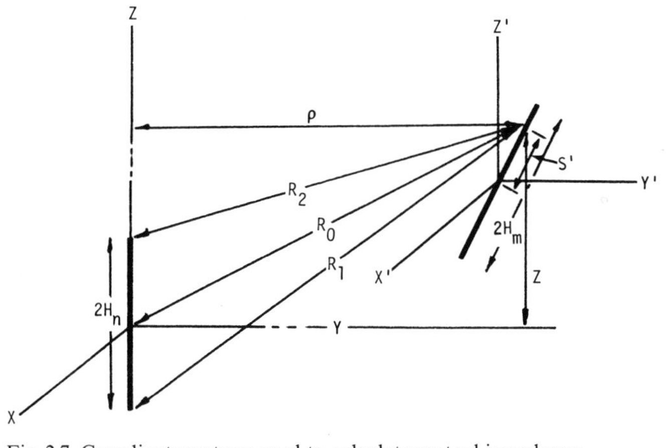

Fig. 2.7. Coordinate systems used to calculate mutual impedance

2.2. Method of Moments Applied to Log-Periodic Dipole Antennas

One of the first computer solutions to the log-periodic dipole antenna (LPDA) was derived by CARRELL [2.1]. The analysis was formulated in terms of impedance and admittance matrices for the dipole and transmission line networks. The mutual impedances between the dipole elements were calculated using the induced EMF method assuming symmetric sinusoidal current distribution on the dipoles.

Carrell, by utilizing the induced EMF method, assumed that the self-impedance of each element is the same when in the array as when isolated and that the effects of changes in the current distributions due to mutual interaction are ignored. These approximations are acceptable for some engineering purposes when the dipoles are thin and a half-wavelength in length or less. They are not adequate for longer or thicker antennas. It so happens, however, that the construction and usual operation of most log-periodic antennas are such that the powers in the elements that are much shorter or longer than half-wavelength are relatively small and their contributions to the overall characteristics of the array are not critical. For this reason Carrell's results have proved adequate in the range of frequencies in which dipoles not too near the ends of the antennas are resonant and therefore lend insight into the operation of LPDA's and were able to be used as a design tool. On the other hand, a complete analysis of the detailed characteristics of the array over a wider range of operating conditions has not been provided.

Cheong and King [2.11, 12] applied their three-term theory to a single LPD antenna and Kyle [2.13] was the first to apply the method of moments as described by Harrington [2.2] to solve the impedance matrix, and extend the computations to more than one LPDA. However, Cheong and King's results are limited to the cases where the maximum dipole lengths are less than one and one-quarter wavelength. Also, some of the approximations used by Kyle in computing the dipole impedance matrix and also in computing the far field pattern yielded a solution that was no more accurate than the original solution given by Carrell. Kyle assumed that each element of the dipole impedance matrix can be obtained by solving a two-dipole mutual impedance problem, where all elements not directly involved are removed from the system. This same approximation was used by Carrell. Kyle obtained the pattern function by assuming that the current distribution on each dipole was as if it were in free space. In addition, Kyle neglected the effect of the feeder booms of the coupled LPDA's.

The present formulation also uses the method of moments in solving the impedance matrix. However, in the moments solution the method of subsectional basis is applied with both the expansion and testing functions being sinusoidal distributions as described by Richmond [2.4] and Imbriale and Ingerson [2.10]. A basic advantage is that one segment per dipole is equivalent to the induced EMF method (Carrell's solution) and hence gives a physically meaningful result. Also, the formulation is generally enough to allow the inclusion of arbitrary wires (not LPDA's) in the vicinity of the LPDA's and can therefore include such effects as supports and feeder booms. This method overcomes some of the difficulties and approximations of the previously proposed methods.

2.2.1. Formulation of the Problem

The geometry of a single log-periodic dipole antenna is illustrated in Fig. 2.8, along with some of the definitions commonly used in its description. The general problem to be solved consists of M arbitrarily placed mutually coupled LPD antennas.

The approach used to set up this problem is a generalization of that originally developed by CARRELL and used later by KYLE. The problem is divided into two parts; calculating the voltages and currents along the feeder lines and calculating the far fields of the radiating dipoles. For the calculation of currents and voltages, the log-periodic dipole antennas are considered as the parallel connection of two N-port networks. One N-port represents the mutual coupling between the dipole elements. The other represents the transmission lines that interconnects the dipoles.

The approach is shown schematically in Fig. 2.9. The N-port labeled "dipole elements" includes the self and mutual impedances between N unconnected dipoles located arbitrarily in space. The "transmission line" N-port represents the transmission line connecting the dipole elements. Included in this network is the effect of reversing the polarity between successive dipoles on each individual antenna.

For the general case of M antennas, each one has a current source I_{SM}. If there are N_e dipole elements on each antenna, and the elements are counted starting from the longest, there are sources on ports N_e, $2N_e$, $3N_e$, etc. Similarly, there is a terminating admittance on each antenna,

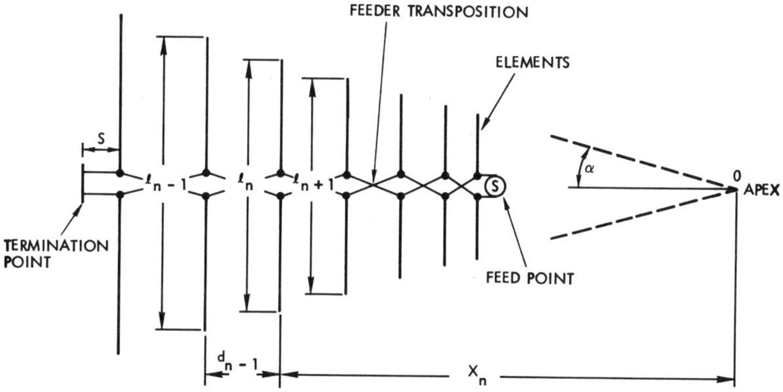

$$\frac{x_n}{x_{n-1}} = \frac{\ell_n}{\ell_{n-1}} = \tau \qquad \frac{d_n}{2\ell_n} = \sigma \qquad h_n = \ell_n/2$$

Fig. 2.8. LPD geometry

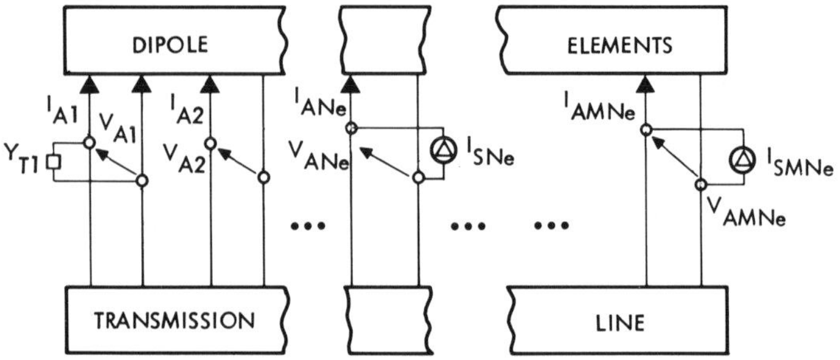

Fig. 2.9. Schematic representation of dipole elements and transmission line

Y_T, and these are located at ports 1, $N_e + 1$, $2N_e + 1$, etc. The N-port networks are of dimensions $M \cdot N_e$.

To write the circuit equations for these networks, let $\boldsymbol{Y}_A$ and $\boldsymbol{Z}_A$ be the short-circuit admittance matrix and open-circuit impedance matrix, respectively, for the "dipole elements" network. Let $\boldsymbol{Y}_F$ be the short-circuit admittance matrix for the "transmission line" feeder network. Let $\boldsymbol{I}_A$ and $\boldsymbol{V}_A$ be the column matrices representing the voltage and current at each port of the "dipole elements" network. Since the two networks are in parallel, the total current $\boldsymbol{I}_S$ can be written as

$$\boldsymbol{I}_S = (\boldsymbol{Y}_A + \boldsymbol{Y}_F)\, \boldsymbol{V}_A\,, \tag{2.32}$$

where $\boldsymbol{I}_S$ represents the applied current sources.

The dipole currents and voltages are given by

$$\begin{aligned} \boldsymbol{V}_A &= (\boldsymbol{Y}_A + \boldsymbol{Y}_F)^{-1}\, \boldsymbol{I}_S \\ \boldsymbol{I}_A &= \boldsymbol{Y}_A \boldsymbol{V}_A\,. \end{aligned} \tag{2.33}$$

The terms of the vectors $\boldsymbol{I}_A$ and $\boldsymbol{V}_A$ represent the input currents and voltages for each dipole element. If the current distribution along the dipole were known it would then be a simple matter to calculate the radiation field. The current distribution will be solved using the method of moments, that is, each dipole element is divided into q_n subsegments and a generalized impedance $[Z]$ is obtained to describe the electromagnetic interactions between the subsections. A matrix equation of the form:

$$[Z]\,[I] = [V] \tag{2.34}$$

is obtained, where $[I]$ is related to the current on subsections, and $[V]$ to the electromagnetic excitation of the subsections. This matrix has a dimension of $P \times P$ where

$$P = \sum_{n=1}^{N} q_n N_e . \tag{2.35}$$

From the solution of the transmission line-dipole elements network equation the voltage excitation consists of the terms of $[V_A]$ applied to the center of each dipole on the LPD antennas and zero elsewhere. From (2.34) the currents on the subsegments are determined and utilizing these currents the far-field radiation pattern is determined. The problem has been formulated and its solution indicated. It remains to determine the explicit forms of the transmission line and dipole elements admittance matrices, and the far field radiation patterns.

The Transmission Line Admittance Matrix

The transmission line admittance matrix, denoted $\boldsymbol{Y}_F$, is composed of M terminated transmission lines with a port at the position where each dipole is connected. Since $\boldsymbol{Y}_F$ is a short-circuit admittance matrix, a given element $(Y_F)_{mn}$ represents the current induced across port m (which is shorted) by a unit voltage at port n, with all other ports shorted. Thus, $(Y_F)_{mn}$ is non-zero only for $n-1<m<n+1$. It is also obvious that there is no current induced at a transmission line port of one antenna by a voltage on another.

For a system composed of M antennas, the form of transmission line admittance matrix is simply

$$\boldsymbol{Y}_F = \begin{bmatrix} Y_{F1} & \mathbf{0} & \mathbf{0} & \ldots \mathbf{0} \\ \mathbf{0} & Y_{F2} & \mathbf{0} & \ldots \mathbf{0} \\ \mathbf{0} & \mathbf{0} & Y_{F3} & \ldots \mathbf{0} \\ & \cdots & & \cdots \\ \mathbf{0} & \mathbf{0} & \mathbf{0} & \ldots Y_{FM} \end{bmatrix}, \tag{2.36}$$

where $\boldsymbol{Y}_{Fi}$ is the transmission line admittance matrix for a single antenna and can be obtained from CARRELL [2.1, p. 25] as

$$\begin{bmatrix} (Y_{Ti}-jY_0\cot kd_1) & -\mathrm{j}\,Y_0\csc kd_1 & 0 & \ldots & 0 \\ -jY_0\csc kd_1 & -jY_0(\cot kd_1+\cot kd_2) & -jY_0\csc kd_2 & \ldots & 0 \\ 0 & -jY_0\csc kd_2 & -jY_0(\cot kd_2+\cot kd_3) & \ldots & 0 \\ \cdots & \cdots & \cdots & \cdots & \cdots \\ 0 & 0 & 0 & \ldots & -jY_0\cot kd_{N_e-1} \end{bmatrix}, \tag{2.37}$$

where Y_0 is the transmission line characteristic admittance, k the propagation constant of the transmission line, and Y_{Ti} accounts for the rear element termination. Account has been taken of the polarity reversal between neighboring ports by reversing the sign of the off-diagonal terms.

The Dipole Elements Admittance Matrix

As stated earlier, $\boldsymbol{Y}_\mathrm{A}$ and $\boldsymbol{Z}_\mathrm{A}$ are the short-circuit admittance matrix and opencircuit impedance matrix, respectively, for the "dipole elements" network. All dipoles are as if they were in free space, with no transmission line connecting them. More precisely, an element of $\boldsymbol{Y}_\mathrm{A}$, $(\boldsymbol{Y}_\mathrm{A})_\mathrm{ab}$, represents the current induced in dipole a, by a voltage on dipole b with all other dipoles short circuited. It should be mentioned at this point that the system may not only consist of the dipoles for the antennas but allows other scattering wires in the vicinity.

The solution will be obtained utilizing the generalized network parameters and multiport system concept as described by HARRINGTON [2.2, Chapter 6].

The elements of $\boldsymbol{Y}_\mathrm{A}$ can be related to electromagnetic field quantities as follows: Let E^a, H^a be the field due to a voltage source v_a applied to port a, all ports short circuited, and M^b, J^b be the current due to v_b applied to port b, all ports short circuited. The admittance matrix element $(Y_\mathrm{A})_\mathrm{ab}$ is then given by

$$(Y_\mathrm{A})_\mathrm{ab} = (v_\mathrm{a} v_\mathrm{b})^{-1} \int (H^\mathrm{a} \cdot M^\mathrm{b} - E^\mathrm{a} \cdot J^\mathrm{b})\, d\tau\,, \tag{2.38}$$

where the integration extends over all space and reduces to

$$(Y_\mathrm{A})_\mathrm{ab} = (v_\mathrm{a} v_\mathrm{b})^{-1} \int E^\mathrm{a} \cdot J^\mathrm{b}\, d\tau \tag{2.39}$$

for the case of non-magnetic media. Equation (2.39) will now be solved using the method of moments.

Each wire element of the "dipole elements" network is divided into subsegments and a generalized impedance matrix $[Z]$ composed of the self and mutual impedances between subsegments calculated using (2.8) is obtained.

The matrix Eq. (2.34) describes the electromagnetic interactions. The total current $\boldsymbol{J}$ is then an equation of the form

$$\boldsymbol{J} = \sum_n I_n S_n\,, \tag{2.40}$$

where S_n is the piecewise sinusoid given by (2.4), and n extends over all the dipoles. If current J^b is due to voltage v_b applied to port b, identify

$[V_a]$ as the generalized voltage vector due to voltage v_a, and write the current for J^b as follows

$$J^b = [S_n][I_n] = [S_n][Y][V_a], \tag{2.41}$$

where $[S_n] = [S_1, S_2, \dots S_n]$ and $[Y] = [Z]^{-1}$.

Substituting (2.41) into (2.39) and identifying $[V_b]$ as the voltage vector due to v_b we have

$$[Y_A]_{ab} = (v_a v_b)^{-1} [V_a][Y][V_b]. \tag{2.42}$$

At this point, it is important to recall that the elements of the impedance matrix $[Z]$ and hence the admittance elements of $[Y]$ are equivalent to the familiar impedances and admittances found using the induced EMF method. If, instead of using a number of segments for each dipole of the LPD antenna, only one segment per dipole was used, the admittance matrix $\boldsymbol{Y}_A$ would be the same as that calculated using the induced EMF method and thus would correspond exactly to the dipole admittance matrix used by CARRELL in his solution of the LPDA. Hence, Carrell's solution can be obtained from the moments solution by simply using one segment per dipole and is therefore referred to as a "first-order" theory.

Far-Field Radiation Patterns

Once the N-port network problem is solved, the next step is to calculate the far field pattern. To compute the far field, it is necessary to have the current distribution over the entire length of each of the dipoles which can be obtained if the vector $[I]$ of (2.34) is known. Since the generalized admittance matrix $[Y]$ is known, the current can be calculated once an exciting voltage vector $[V]$ is determined. Solving the LPD N-port problem gives the voltages V_A and currents I_A at the dipole terminals. The voltages V_A can then be considered as the voltage sources exciting the array of dipoles that make up the antenna. It is then straightforward to obtain the current $[I]$ and calculate the far field.

For the case of sinusoidal expansion functions, the calculation of the antenna pattern is quite easy. The pattern results from the superposition of the fields of the many small subsegments with sinusoidal current distributions. The far-field of a single small arbitrarily located dipole of this type with half-length H is (see Fig. 2.10a)

$$\begin{aligned} \boldsymbol{E}(\theta, \varphi) = j\eta(2\pi r)^{-1} \exp(-jkr)(\hat{\boldsymbol{u}}_r \cdot \hat{\boldsymbol{s}}) \exp(j\hat{\boldsymbol{k}}_r \cdot \hat{\boldsymbol{r}}_0) \\ \cdot [\cos(kH\cos\zeta) - \cos kH] I_0/(1-\cos^2\zeta), \end{aligned} \tag{2.43}$$

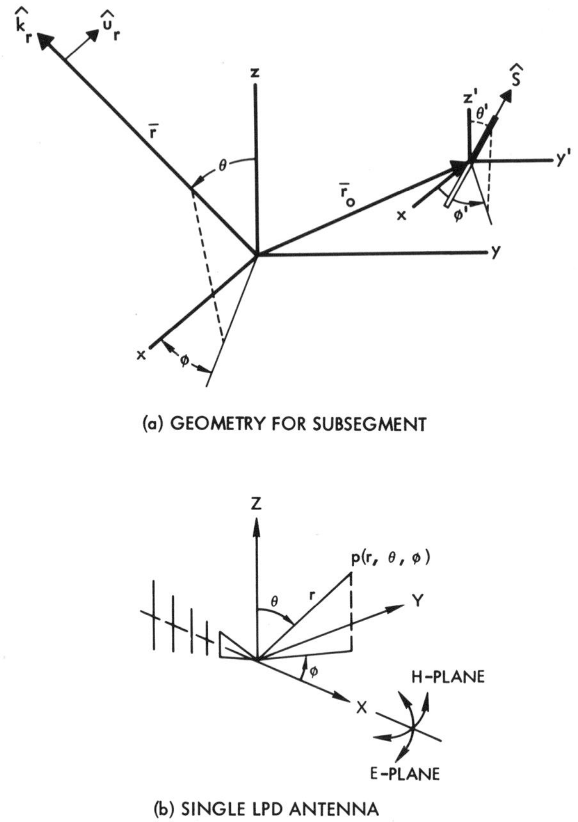

Fig. 2.10a and b. Geometry for calculation of the far field

where (x_0, y_0, z_0) is the position of the dipole, θ', ϕ' its angular rotation, $\hat{\boldsymbol{u}}_r$, $\hat{\boldsymbol{s}}$, $\hat{\boldsymbol{k}}_r$ are unit vectors, ζ the angle between $\hat{\boldsymbol{k}}_r$ and $\hat{\boldsymbol{s}}$, I_0 the current amplitude, η the impedance of free space and $k = 2\pi/\lambda$ is the propagation constant of free space. The gain of the system of antennas for polarization $\hat{\boldsymbol{u}}_r$ may be defined as

$$g(\theta, \phi) = 4\pi r_0^2 \eta^{-1} |\boldsymbol{E}(\theta, \phi)|^2 / P_{\text{in}}, \tag{2.44}$$

where $\boldsymbol{E}$ is now the total field from all the antennas in the system, i.e., the sum over all the subsegments of each dipole and of all the dipoles in the arrays and P_{in} is the total input power which can be calculated as $P_{\text{in}} = \operatorname{Re}\{[\tilde{V}][I^*]\}$.

For the case of only one antenna and with one subsegment per dipole (first-order theory) the radiation patterns are given by

$$|\boldsymbol{E}| = \sum_{i=1}^{N_e} I_{ai} \exp[-jk|X_i| \sin\theta \cos\phi] \cdot [\cos(kh_i \cos\theta) - \cos kh_i]/(\sin\theta \sin kh_i)| , \tag{2.45}$$

where (θ, ϕ) are given in Fig. 2.10b, h_i, x_i are defined in Fig. 2.8 and I_{ai} are the elements of the dipole current matrix. Equation (2.45) is presented because it is in the radiation-pattern calculation that CARRELL has an error. Instead of the $\sin\theta$ term that appears in the denominator, CARRELL incorrectly has this term in the numerator which leads to E-plane patterns which are narrower than should be. Thus, in the examples given by CARRELL the measured E-plane patterns were consistantly broader than the calculated. In the examples that follow, it will be shown that the first-order theory gives a very reasonable representation of the LPD antennas performance.

2.2.2. Single Log-Periodic Dipole Antenna

The accuracy of the analysis and computer program was verified with experimental results and by comparisons with previously published

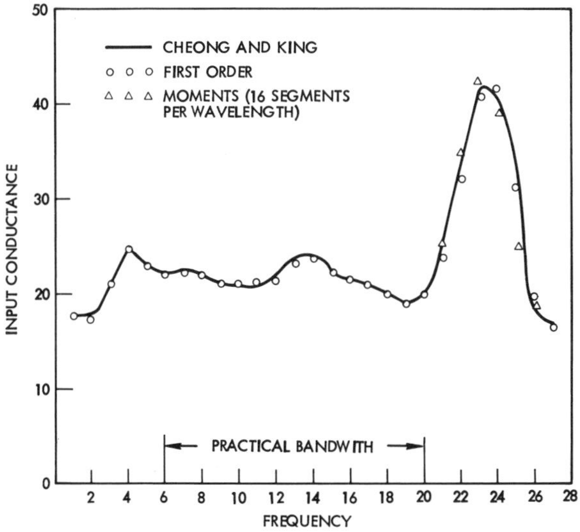

Fig. 2.11. Input conductance in millimhos for antenna No. 1

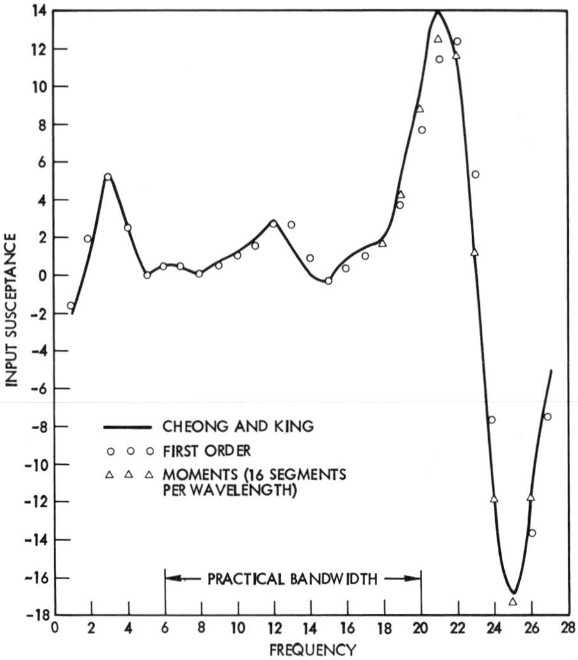

Fig. 2.12. Input susceptance in millimhos for antenna No. 1

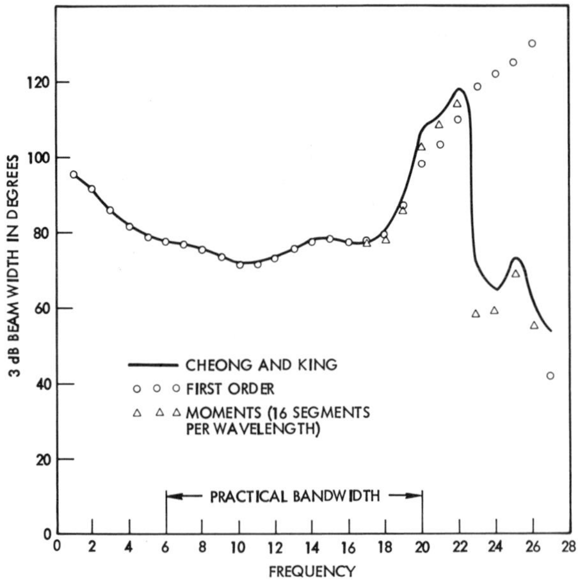

Fig. 2.13. H-plane half-power beamwidths for antenna No. 1

data. An attempt is also made to determine when the "first-order" solution [2.1] is adequate and when the entire moments solution of the antenna is required.

Figures 2.11–13 compare the input admittance and H-plane patterns of CHEONG, and King's 12 element antenna [2.11] for the first-order, moments and their three-term theory. The LPD is designated as Antenna No. 1, the definition of the parameters is given in Table 2.1, and the frequencies in [2.11, p. 1317]. The first-order solution compared quite favorably with the three-term theory. For the frequencies calculated, the dipoles ranged in length from 0.2 to 1.2 wavelengths. Therefore, the length of the predominance of dipoles was less than one wavelength, and since the induced EMF method for calculating impedances is known to be reasonably accurate for lengths less than one wavelength, it is not surprising that Carrell's solution gives good results. Although

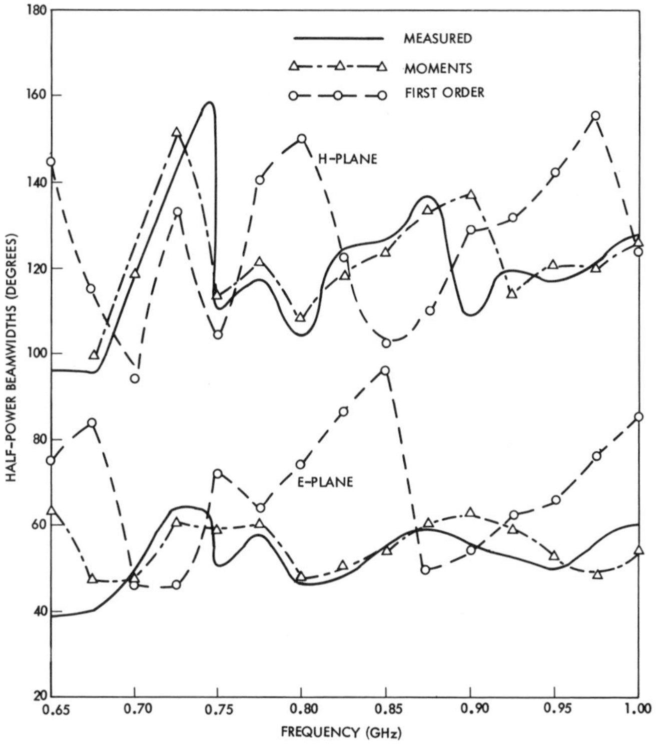

Fig. 2.14. Measured and calculated half-power beamwidths versus frequency for antenna No. 2

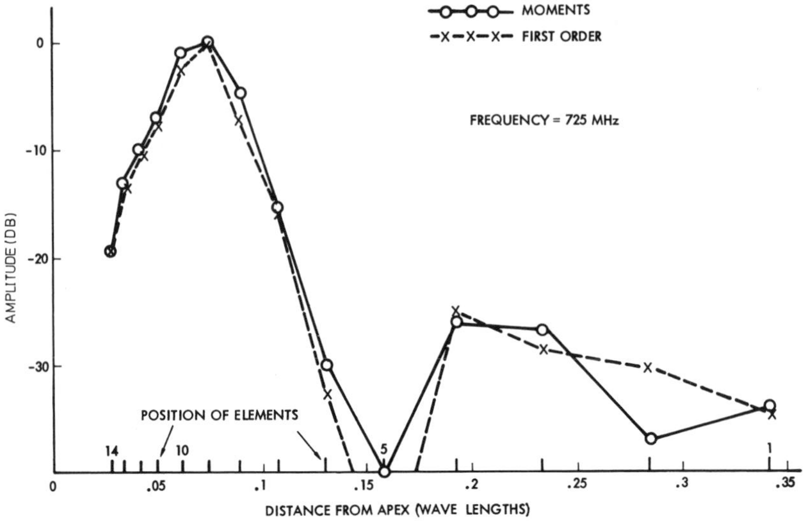

Fig. 2.15. Computed amplitude of dipole element base current versus relative distance from the apex for antenna No. 2

not specifically displayed in these figures, the first order and moments calculations agreed for frequencies less than f_{18}.

As a second example, consider the data shown in Fig. 2.14 for antenna No. 2, where a slightly more compressed array (larger α angle) is used with the *longest dipole* being 1.87 wavelengths at the lowest frequency, 0.65 GHz, to 2.89 wavelengths at the highest frequency 1.0 GHz. Here the moments solution is clearly superior to the first-order solution of CARRELL. Observe that CHEONG and King's three-term theory cannot be applied to this case as it is limited to dipoles of lengths less than 1.25 wavelengths.

For antenna No. 2 the calculated dipole currents at 725 MHz for both "first-order" and moments solution (10 segments per wavelength) are displayed in Fig. 2.15. Both solutions yield similar dipole currents; however, the H-plane patterns are quite different (Fig. 2.16). Also shown on the graph is a pattern obtained utilizing the dipole currents calculated via the moments technique and assuming that the current distribution on the dipoles was purely sinusoidal as in the first-order theory. The observation that this calculation more nearly corresponds to the first order solution than the moments solution indicates that the actual current distribution of the in-place dipole elements is important and the assumption made by KYLE of "solving for the currents on each as if it were in free space" does not seem justified.

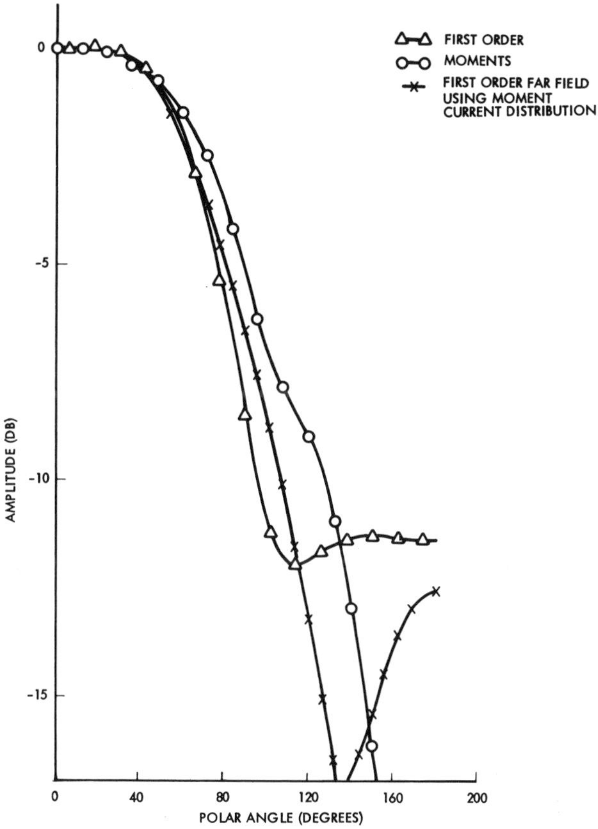

Fig. 2.16. H-plane radiation pattern for antenna No. 2 at 725 MHz

The discrepencies between the two solutions can be traced to the effects of energy leaking past the first active region and exciting the 3/2 wavelength elements. The first-order theory does not correctly characterize the current distribution on the 3/2 wavelength element and because it is of resonant length it can make a contribution to the radiation pattern when excited. Additional information including calculated current distributions and comparisons to measured patterns can be found in [2.14].

2.2.3. Coupled LPD Antennas

Several computations were carried our for systems of two, three and five log-periodic antennas for the geometry displayed in Fig. 2.17. The

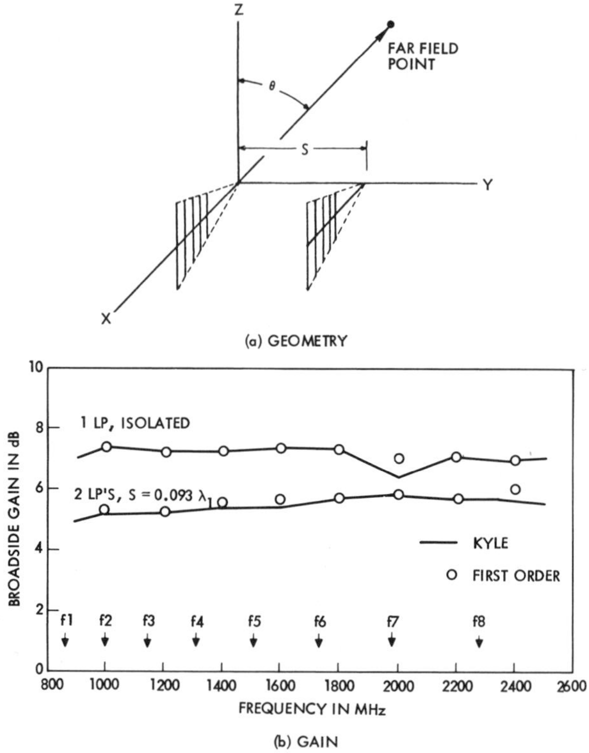

Fig. 2.17a and b. Broadside gain of in-place element pattern versus frequency for antenna No. 3

far-field patterns were calculated as a function of frequency and the spacing *S*. Comparisons are made between feeding one antenna alone and feeding one antenna with the others nearby.

The broadside gain versus frequency for a two-element array with one antenna fed is shown in Fig. 2.17 for antenna No. 3. Compared are the calculated results of KYLE to the "first-order" theory. Since the dipoles are generally smaller than one wavelength, it is not necessary to use the complete moments solution. KYLE concluded from his studies that the broadside gain of the in-place element pattern for a fixed spacing is relatively flat versus frequency; however, this conclusion does not seem to be borne out by the measured and calculated data for antenna No. 4 presented in Fig. 2.18, where the difference in gain between feeding one antenna of a two-element array and a single LPDA are presented. Figure 2.19 is a calculation of the broadside gain of antenna No. 4 when surrounded by two additional antennas and four additional an-

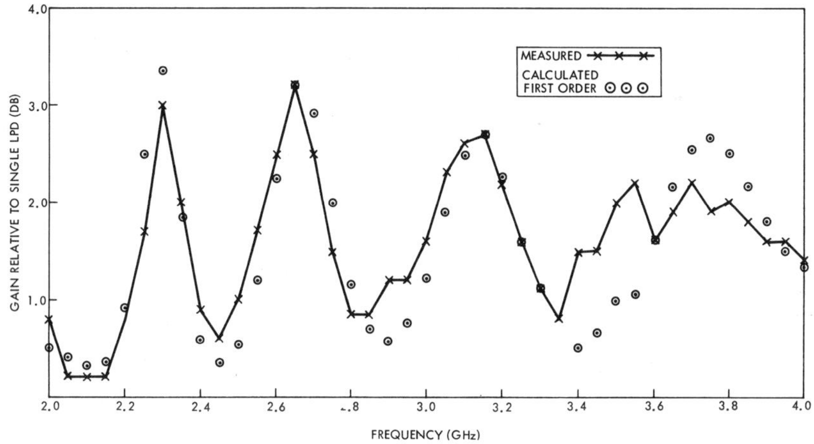

Fig. 2.18. Broadside gain relative to a single LPDA for a two-element array of antenna No. 4

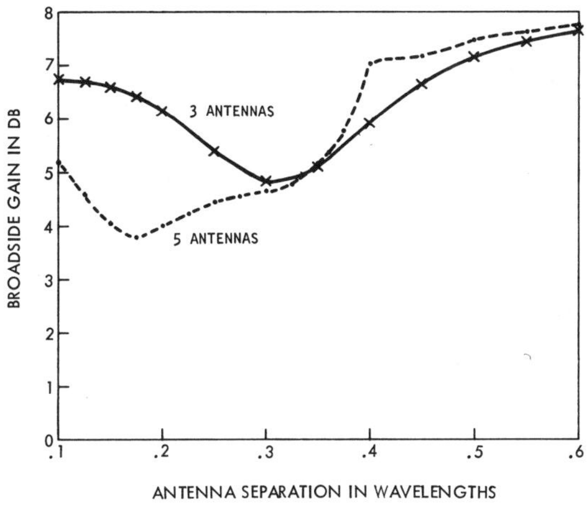

Fig. 2.19. Broadside gain of center LPDA in a 3-antenna and 5-antenna array

tennas as a function of antenna separation in wavelengths, and illustrates that mutual coupling is significant even for spacings as far apart as 1/2 wavelength.

2.2.4. Effects of Feeder Boom

In the previous examples when the mutual impedances were calculated, only the dipole elements themselves were considered, and the feeder boom which forms the transmission line was not considered. For a

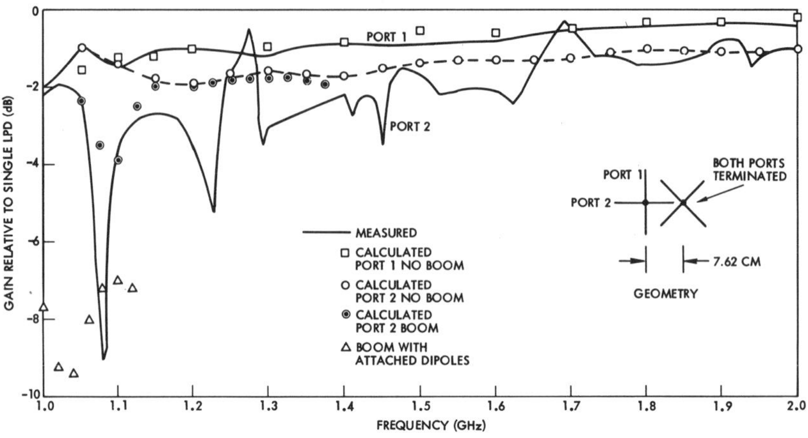

Fig. 2.20. Measured and calculated effects of feeder boom

single antenna there is no problem since the feeder boom is directed through the center of the dipoles and is therefore theoretically decoupled from the dipoles. In fact, any dipoles in the bisector plane are decoupled and thus for the examples given by KYLE and in Subsection 2.3.3 there is no boom coupling. However, when there is more than one antenna, the feeder boom of the second LPDA, if it is not along the center line of the dipoles of the first, can couple to the dipoles and its effect should be considered in the calculation.

As a rather extreme example consider the geometry and data presented in Fig. 2.20. There are two crossed dipoles with center to center separations of 7.62 cm. Only one port of one of the LPDA's is excited, the others are terminated. For Port 1, the feeder boom of the antenna is along the bisection of the dipoles and there is no coupling. For Port 2, the feeder boom can interact and the measured results show a dramatic difference. Calculations made with and without considering the boom interaction are shown. The feeder booms were modeled as a straight wire from the tip of the antennas to the rear termination and was not connected to the dipoles. A first-order solution was used on the LPDA's with 16 segments per wavelength on the booms. A moments solution on the LPDA did not significantly alter the results. Also shown in Fig. 2.20 is a calculation made by modeling the second LPDA as a straight wire boom with attached dipoles. This modeling does not take into account the effects of interconnecting the dipoles with a transmission line and is therefore inadequate.

Additional calculations and measurements were made to verify whether or not the "unattached boom model" successfully predicted gain perturbations caused by interaction with the boom. The results indicated, that for less severe coupling (losses less than a few dB) the model was adequate (see [2.14]).

Based upon calculated and measured results, it is possible to draw certain conclusions regarding the effects of mutual coupling in log-periodic dipole arrays:

1) The effects of feeder boom are significant for closely spaced arrays.
2) The mutual coupling effects are oscillatory when measured as a function of frequency.
3) There is not a monotonic trend in the mutual coupling as the separation between the antennas is decreased with the frequency held constant and there are significant effects for spacings as large as 0.5 λ.

2.2.5. The Log-Periodic Fed Yagi

The Yagi-Uda array [2.15, 16] consists of a driven element plus a number of parasitic elements that increase the gain or directivity of the radiation pattern over that of a dipole antenna. The number of parasitic elements, their length and their spacing with respect to the driven element determine the characteristics of the array.

Generally speaking, the Yagi provides a greater gain than a LPD antenna, but is somewhat restricted in bandwidth.

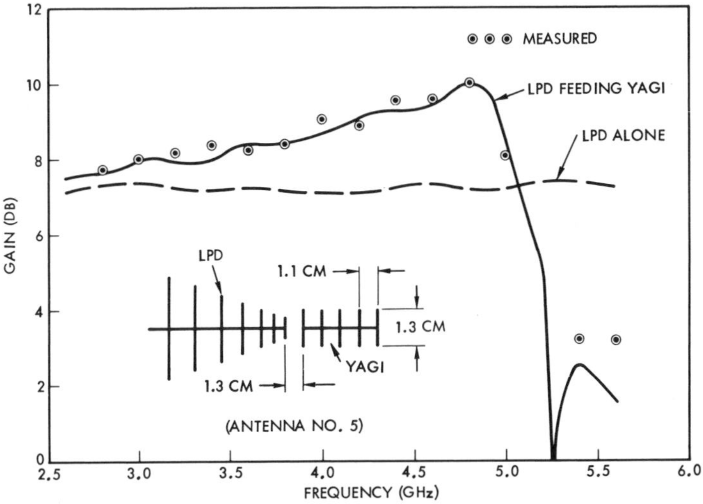

Fig. 2.21. Measured and calculated gain for the log-periodic fed Yagi

In this subsection we wish to display another example of the versitility of the theory presented in Subsection 2.2.1 as well as demonstrate a technique for broadbanding the Yagi.

The Yagi functions as a directional antenna having power gain by virtue of the proper phasing of the director elements as well as the proper excitation of these elements by the driven-reflector element interaction (see THIELE [2.17].) The Yagi can be broadbanded by utilizing a LPD antenna to feed the parasitic elements instead of a single dipole. This arrangement gives greater gain than the LPDA alone and provides greater bandwidth than the Yagi alone. An example is provided in Fig. 2.21, where antenna No. 5 is used to feed a 5 element Yagi. As shown, the LPD Yagi combination provides greater gain than the LPD antenna alone for an octave bandwidth.

2.3. Wire Antennas as Feeds for Parabolic Reflectors

Numerical techniques are extremely useful for designing systems of single or multiple log-periodic dipole antennas used as feeds for a parabolic reflector. The LPDA geometry readily lends itself to an accurate computer solution of its gain, voltage standing wave ratio (VSWR), and radiation patterns and once the primary patterns of the LPDA's are calculated it is a straightforward procedure to incorporate these patterns into the calculation of the reflector secondary radiation patterns and gain.

The LPD antenna is attractive as a broadband feed for parabolic reflectors despite its tendency to cause defocusing because of axial phase-center movement with frequency. The gain loss due to defocusing in a parabolic reflector is only a function of the defocusing in terms of wavelengths and not dependent on the diameter of the reflector. Hence, antennas such as the LPDA which have a frequency independent phase center location in terms of wavelengths from some geometrical point (virtual apex) will experience a frequency independent defocusing loss over its bandwidth if the virtual apex is placed at the focal point. The optimum feed design for maximum gain will change with varying reflector f/D (reflector focal length to diameter ratio) since spillover, aperture illumination, and defocusing loss are functions of f/D, while primary pattern beamwidth and symmetry, gain and VSWR are functions of the feed parameters.

If the LPD antenna is to be used for narrowband operation, the phase center is placed at the focal point and the defocusing loss is eliminated. It is to be expected that the feed designs which optimize gain will be different for narrowband and broadband cases.

In Subsection 2.3.3 a technique for reducing the mutual coupling between LPD antennas is presented that is useful when multiple feeds are used.

2.3.1. Scattering from Parabolic Reflectors

To compute the scattered field from a parabolic reflector the physical-optics technique has been adapted. Two general formulations are presented. A complete vector formulation which is required if the primary field has no symmetry and a simpler scalar formulation which can be utilized if the primary field can be adequately approximated by E- and H-plane patterns. The analysis parallels that given in [2.18] so only the pertinent details are included.

The far-field secondary pattern for a parabolic reflector is given by

$$E(P) = -j\omega\mu(4\pi R)^{-1}\exp(-jkR)\int_{\text{surface}}[\boldsymbol{J}_s - (\boldsymbol{J}_s\cdot\hat{\boldsymbol{u}}_R)\,\hat{\boldsymbol{u}}_R]\cdot\exp(jk\boldsymbol{\varrho}\cdot\hat{\boldsymbol{u}}_R)\,dS\,, \tag{2.46}$$

where $\boldsymbol{J}_s$ is the reflector surface current, $k = 2\pi/\lambda$, $\hat{\boldsymbol{u}}_R$ the unit position vector of the far-field point, dS the incremental surface area, and the quantities R, ϱ and P as defined in Fig. 2.22. The physical-optics ap-

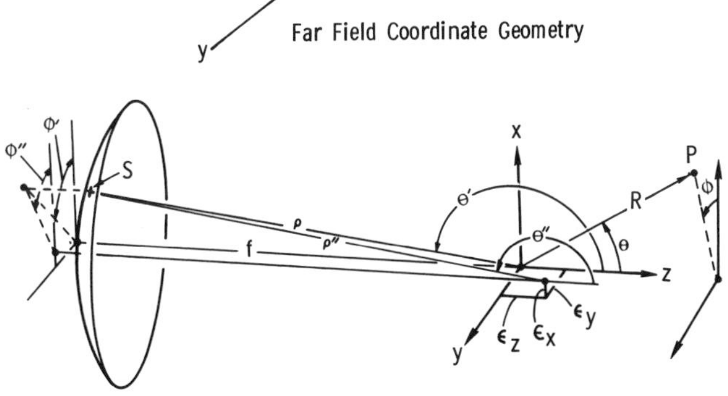

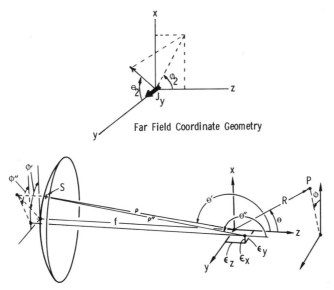

Fig. 2.22. Reflector coordinate system

proximation assumes that the incident field $H_{\rm inc}$ from the source is reflected optically which gives the currents $\boldsymbol{J}_s = 2\hat{\boldsymbol{n}} \times \boldsymbol{H}_{\rm inc}$, where $\hat{\boldsymbol{n}}$ is the outward normal from the surface. The most general form of the incident field in the double primed coordinate system is as follows

$$\begin{aligned} \boldsymbol{E}_{\rm inc} &= (\varrho'')^{-1}[\exp(-jk\varrho'')]\,[F_1(\theta'',\phi'')\,\hat{\boldsymbol{u}}_{\theta''} + F_2(\theta'',\phi'')\,\hat{\boldsymbol{u}}_{\phi''}] \\ \boldsymbol{H}_{\rm inc} &= (\hat{\boldsymbol{u}}_{\varrho''} \times \boldsymbol{E}_{\rm inc})/\eta_0\,. \end{aligned} \tag{2.47}$$

For a complete vector solution of the problem, the incident fields are determined by a moments solution of the primary feed geometry. In particular, the quantities $F_1(\theta'',\phi'')$ and $F_2(\theta'',\phi'')$ which are the $\hat{\boldsymbol{u}}_{\theta''}$ and $\hat{\boldsymbol{u}}_{\phi''}$ components of the far field are computed under the assumption of unity input power. The reflector is assumed to be in the far field of the feed geometry and that there is no interaction with the scattered energy. A two-dimensional integration of (2.46) gives the far-field secondary pattern with the gain of the total antenna system calculated by the definition:

$$G(\theta,\varphi) = 4\pi \cdot \text{Radiated power per unit solid angle.}$$

For the scalar calculation, the incident field is approximated in the following manner

$$\boldsymbol{E}_{\rm inc} = (\varrho'')^{-1}\exp(-jk\varrho'')\,[E_{\rm p}(\theta')\sin\phi'\,\hat{\boldsymbol{u}}_{\theta'} - H_{\rm p}(\theta')\cos\varphi'\,\hat{\boldsymbol{u}}_{\phi'}]\,. \tag{2.48}$$

The surface current is again obtained using the physical-optics approximation where $E_{\rm p}(\theta')$ is the E-plane radiation pattern and $H_{\rm p}(\theta')$ is the H-plane radiation pattern.

There will, in general, be x, y, and z directed current components. For the scalar case, which assumes fields near the boresight, we ignore the z and x directed currents so that the secondary radiation field has a principle polarization vector component ($\hat{\boldsymbol{u}}_{\theta_2}$ directed) in the (θ_2,ϕ_2) coordinate system of Fig. 2.22.

In the vector formulation, when the feed is laterally defocused, the illumination function is assumed stationary with respect to the food and ϱ'' is obtained exactly using the law of cosines. In the scalar formulation, the illumination function is assumed stationary with respect to the reflector and lateral feed displacement is included by approximating ϱ'' by parallel ray theory, as follows. Let $\boldsymbol{\varrho}'' = \boldsymbol{\varrho} - \boldsymbol{d}$, where the feed displacement $\boldsymbol{d} = \varepsilon_x\hat{\boldsymbol{u}}_x = \varepsilon_y\hat{\boldsymbol{u}}_y = \varepsilon_z\hat{\boldsymbol{u}}_z$ whence, if we assume that $(d/\varrho) \ll 1$, we obtain

$$\varrho'' \approx \varrho - (\varepsilon_x \sin\theta'\cos\phi' + \varepsilon_y \sin\theta'\sin\phi' + \varepsilon_z\cos\theta')\,. \tag{2.49}$$

Further, we make the assumption that the observation point is near the boresight so that $\cos\theta \approx 1$ and $\sin\theta \ll 1$. With these approximations (2.46), can be shown to reduce to

$$E(P) = -j(4\pi R)^{-1} \exp(-jkR - j2f) \cdot k(a^2/f) \int_0^1 \exp(jk\varepsilon_z \cos\theta')\, I(r)\, (1-\cos\theta')\, r\, dr\,, \tag{2.50}$$

where

$$I(r) = \int_0^{2\pi} [\sin^2\phi'\, E_p(\theta') + \cos^2\phi'\, H_p(\theta')] \cdot \exp(jk(\varepsilon_x \sin\theta' \cos\varphi' + \varepsilon_y \sin\theta' \sin\phi') + jur\cos(\phi - \phi'))\, d\phi'.$$

Now, making use of some trigonometric manipulations similar to those employed by RUZE [2.19], incorporating the plane-wave to cylindrical-wave transformation and using the following definitions

$$\omega = [(ur)^2 + 2k\varepsilon_R \sin\theta' ur \cos(\phi - \xi) + (k\varepsilon_R \sin\theta')^2]^{1/2}$$

$$\alpha = \tan^{-1}[(ur\sin\phi + k\varepsilon_R \sin\theta' \sin\xi)/(ur\cos\phi + k\varepsilon_R \sin\theta' \cos\xi)]$$

$$\varepsilon_R = \sqrt{\varepsilon_x^2 + \varepsilon_y^2}$$

$$\xi = \tan^{-1}(\varepsilon_y/\varepsilon_x)$$

we obtain the final result

$$E(P) = -jR^{-1} \exp(-jkR - j2f)(\pi D/\lambda)(D/4f) \cdot \int_0^1 (1-\cos\theta') \exp(jk\varepsilon_z \cos\theta') \cdot [J_0(\omega)(E_p + H_p)/2 + (E_p - H_p)\, J_2(\omega) \cos 2\alpha/2]\, r\, dr\,. \tag{2.51}$$

Equation (2.51) can be interpreted as the principal polarization component in the (θ_2, ϕ_2) coordinate system utilizing approximations for the incident field, ϱ'', and some small angle assumptions.

2.3.2. Optimum Design of LPD Feeds

The object of this subsection is to utilize the numerical solutions to develop design data for optimizing the gain of parabolic reflectors fed with LPDA's operating over broad or narrow bandwidths.

The analysis is performed utilizing the previously described techniques to compute primary feed patterns. These primary patterns are then used as inputs for the parabolic scattering analysis to compute the secondary far-field radiation. When this procedure is followed, well behaved performance curves as a function of the basic parameters are found. These curves, when expressed as a function of antenna angle α, scale factor τ, and reflector f/D have well defined optimum.

The selection of the remaining defining parameters can be narrowed to a small range where final selection is flexible, governed mainly by the choice of characteristic impedance which may be acceptable, even though the system gains are the same. These curves then allow the defining parameters of high-gain broad- and narrowband systems to be selected once the reflector f/D is chosen, or, the optimum f/D can be selected.

The parametric curves obtained were based on the first-order solution and the scalar physical-optics scattering from the reflector. Equation (2.45) was used to obtain the E- and H-plane patterns. In the process, however, the results of many computations have been checked using the method of moments solution for the dipole elements. The results were that in midband and for the majority of τ, α, and σ parameters essentially no important differences are found using the higher-order solution. However, for $\tau > 0.9$ and $\alpha < 20°$ the moments solution differs somewhat from the first-order solution and was therefore utilized in obtaining design curves.

The accuracy of the combined primary feed and secondary scattering analysis are verified by correspondence between calculated and experimental measurements of secondary patterns and gain for a parabolic reflector with an $f/D = 0.4$ over the bandwidth of 1–7 GHz.

Design Considerations

In general, the gain of LPD antennas depends on the parameters τ, σ, and α tending towards higher gains, with large τ and σ and small α.

Recalling that the gain/loss ratio due to defocusing in a parabolic reflector for a given f/D is only a function of the defocusing in terms of wavelength and not dependent on the diameter of the reflector, antennas which have a frequency independent phase center location in terms of wavelengths from some geometrical point (virtual apex) which represents the extension of the structure to an infinitely high frequency will experience a constant frequency independent loss over its bandwidth if this point is placed at the focus. Defocusing loss increases with distance. A typical value for a $0.5\,\lambda$ defocusing and an f/D of 0.4 would be about 1.0 dB. However, the smaller the α, the larger the displacement of the phase center from the virtual apex and therefore a larger defocusing loss. Hence, since loss due to defocusing with small α angles is more severe than the aperture distribution losses, antennas with larger α angles are favored for broadband design.

In the practical case, the virtual apex of the antenna is placed ahead of the focal point of the reflector when only a finite bandwidth is being considered. The feed can then be placed so the lowest and highest frequencies have the same defocusing and the rest of the frequency band has less. It is clear that the smaller the bandwidth, the less important is this defocusing loss. For generating design curves, however, two separate cases will be examined. 1) Infinite bandwidth where the virtual apex is placed at the focal point and the defocusing loss is included and 2) zero bandwith—where a single frequency is considered and the feed positioned so that there is no defocusing loss.

It is well known that for a given primary illumination function there is a focal length to diameter ratio (f/D) for which the gain is a maximum. Also, the effect of defocusing is different for varying f/D. Hence, it is to be expected that there is a different LPD parameter selection which will optimize the reflector gain for each f/D.

The design procedure is straightforward. The parameters of an LPDA were studied as a function of secondary on-axis gain. Because the LPD antenna has cyclic gain perturbations as well as front and rear-end truncation effects, a log-cycle in the center of the antenna bandwidth was chosen for the parameter study. The number of dipoles was also chosen large enough to include elements larger than 3/2 wavelengths such that the effects of energy leaking past the first active region would be included. It is desirable to derive design curves that provide stable performance over the whole bandwidth. However, while some parameters actually will produce a higher gain over a portion of each log-period, they may have gain dips over some other portion of each period, and are thus not acceptable.

Several results of this analysis are worth particular mention. First, the half-power beam-width variations of an LPDA are not necessarily a good indicator of the secondary gain stability. That is, while for some parameters the half-power primary beamwidths showed very large excursions, the secondary gain variations were no larger than for other parameter ranges where the primary pattern beamwidths were more stable. This is illustrated for the infinite bandwidth case in Figs. 2.23–26.

In Fig. 2.23 the half-power E- and H-plane beamwidths of an LPDA are plotted over a log periodic of frequency in the center of its bandwidth, as a function of σ with a characteristic feed line impedance $Z_0 = 150\,\Omega$. It is seen that the beamwidths exhibit a large variation over a log period. In Fig. 2.24 the secondary on-axis gains are shown also over the same log period as a function of σ.

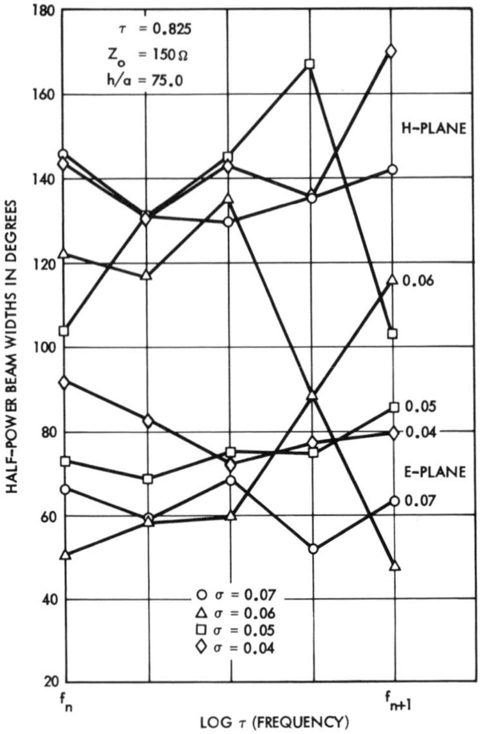

Fig. 2.23. E and H-plane half-power beamwidths over a log-period of frequency as a function of σ with $Z_0 = 150\ \Omega$

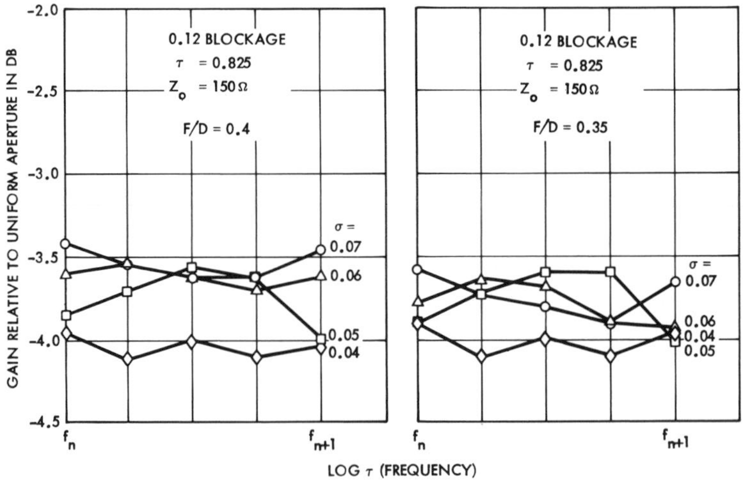

Fig. 2.24. On-axis gain of a parabolic reflector fed with an LPDA over a log-period of frequency as a function of σ with $Z_0 = 150\ \Omega$

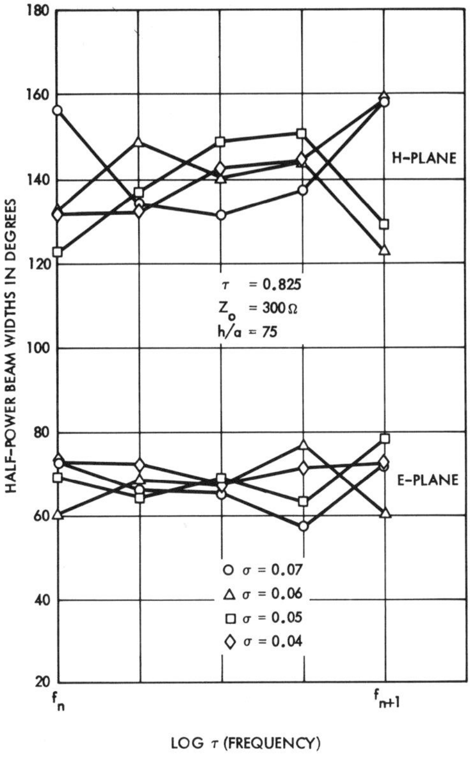

Fig. 2.25. *E*- and *H*-plane half-power beamwidths over a log-period of frequency as a function of σ with $Z_0 = 300\ \Omega$

In Fig. 2.25 the half-power beamwidths of the same parameters over the same log period are shown, but the characteristic feed impedance is $Z_0 = 300\ \Omega$. Comparing Figs. 2.23 and 2.25, the half-power beamwidths are much more stable; however, in Fig. 2.26, which shows the corresponding secondary gain, it is seen that essentially no increase in gain stability was obtained. A possible explanation is that a large part of the beamwidth variation is due to energy that has leaked past the normal active region and excited the 3/2 mode elements. These elements, however, show a much larger defocusing loss and hence their contribution to the secondary gain is markedly reduced.

While it is desirable to have a more stable primary pattern, it is much harder to obtain $Z_0 = 300\ \Omega$ than $Z_0 = 150\ \Omega$, so construction techniques tend to make the lower values more attractive. It is important to mention that if the value of Z_0 is chosen too low, say less than $100\ \Omega$,

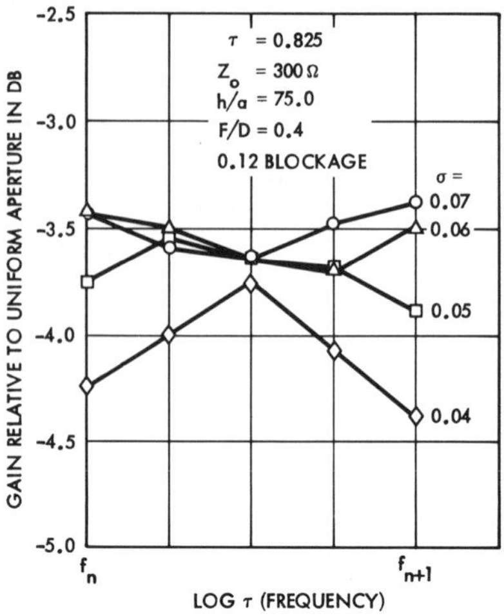

Fig. 2.26. On-axis gain of a parabolic reflector fed with an LPDA over a log-period of frequency as a function of σ with $Z_0 = 300\ \Omega$

for large α angles and low τ serious gain anomalies occur at some portion of the log-periodic [2.20, 21]. A characteristic feed line impedance of $Z_0 = 150\ \Omega$ is generally sufficient to overcome this problem.

Optimum Design

Of prime importance is the finding of the major design parameters σ, τ, and α. Of course, only two of the parameters are independent, the third being related to the other two by $\tan\alpha = (1-\tau)/4\sigma$. The technique that led to the design curves is discussed and then the design curves are presented. Subsequently, the selection of Z_0 is discussed.

For a given f/D, curves were plotted of secondary gain (relative to uniform aperture gain) versus α with τ as a parameter. For values of $\sigma > 0.03$ there was a value of α, independent of τ, which gave the highest gain (see [2.22] for an example). These curves were made for each f/D and the optimum value of α versus f/D was obtained, as shown in Fig. 2.27.

What is needed next is to optimize the τ or σ choice. In general, for a fixed α there is a small increase in directivity with increasing τ. However, for fixed α, increasing τ gives a smaller σ and the VSWR with respect to the mean impedance increases with decreasing σ. Therefore, if the VSWR is to be kept small, there is a lower bound on acceptable σ. For reasonable parameter ranges of τ ($0.8 \leqq \tau \leqq 0.95$) this lower bound is considered to be $\sigma \simeq 0.04$. Below this value of σ the VSWR is generally $\geqq 2:1$ and the

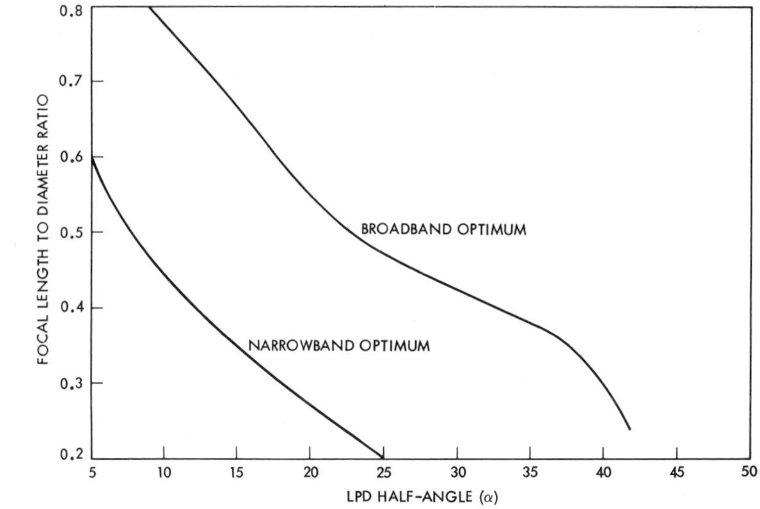

Fig. 2.27. Optimum LPD half-angle α versus f/D

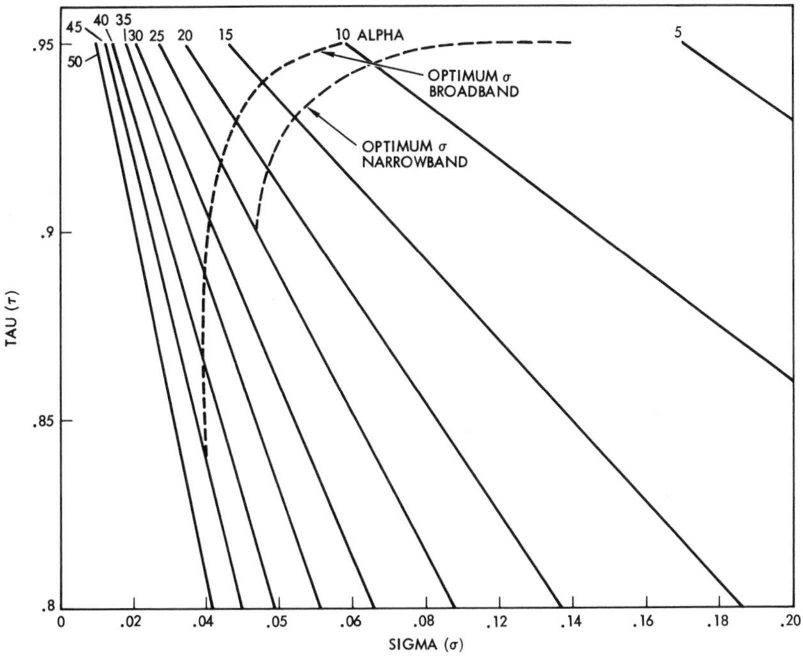

Fig. 2.28. Optimum values of σ for maximum secondary gain

pattern performance unstable. Of course, the acceptable lower bound is also a function of the characteristic impedance Z_0.

The best value of σ or τ can be obtained from Fig. 4.28, where σ versus τ with α as a parameter is shown. For the infinite bandwidth case the optimum values of σ are from $\sigma = 0.04$ to 0.06 with the best choice of σ depending on the exact h/a and Z_0 desired. For the optimum α, σ changes in this range cause only small changes in the secondary gain with the gain going down as σ increases. In general, the exact selection of σ will depend on Z_0. Since low values of $Z_0 (< 150\ \Omega)$ and σ tend toward unstable designs higher σ values should be selected for lower Z_0.

Once the major design parameters σ, τ, and α are selected, the other factors influencing the design of the LPDA are selected, as described by CARRELL [2.1].

Example Case

As an example, suppose we wish to design an LPD which is optimum for $f/D = 0.4$, and is to be matched to a $100\ \Omega$ input line and operates over a 10:1 frequency range. With reference to Fig. 2.27 α is selected as 33°. If we use half-length to radius ratio of 100 for construction ease, then from [2.1, p. 153], $Z_0 = 200$ for a $\sigma = 0.04$ so the optimum value of σ is selected. Tau is obtained from Fig. 2.28 and is 0.9. The bandwidth of the active region from [2.1, p. 147] is 1.22 and the structure band-

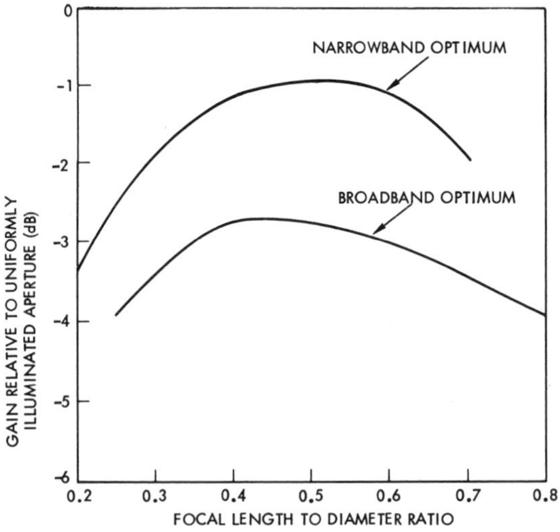

Fig. 2.29. Maximum achievable gain versus f/D ratio

width is 12.2. Therefore, if the longest element is λ max/2, then 24 dipole elements are required. If the virtual apex is placed at the focal point of the reflector, then the gain relative to a uniformly illuminated aperture is -2.8 dB.

If LPD antennas are designed to be optimum for other f/D's the maximum achievable gains are given in Fig. 2.29. From Fig. 2.29 it is seen that the LPD is best for $f/D = 0.4$ to 0.5.

In the following some experimental measurements are shown verifying the secondary gain and pattern predictions using the combined LPD array analysis and the physical-optics scattering.

Experimental Correlation with Theory

Much work has been done to verify that the analysis of CARRELL yields accurate results and that scalar aperture theory yields acceptable performance predictions for high-gain parabolic reflectors. So it is not surprising that (2.51) predicts fairly accurately the performance of an LPDA feeding a paraboloid. A test was conducted utilizing antenna No. 5 placed in a 2.74 m $f/D = 0.4$ parabolic reflector with the virtual

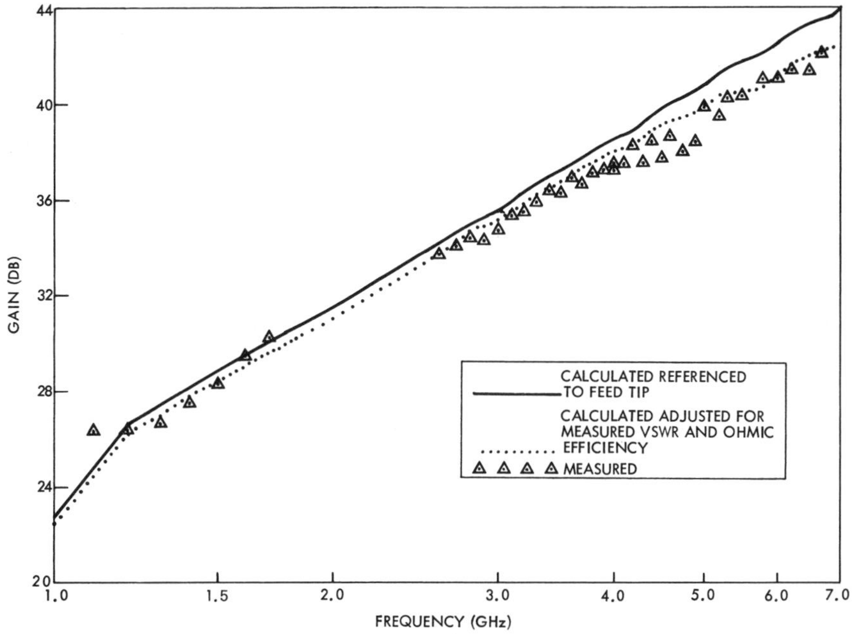

Fig. 2.30. Measured and calculated gain for antenna No. 5 feeding a 2.74 m $f/D = 0.4$ parabolic reflector

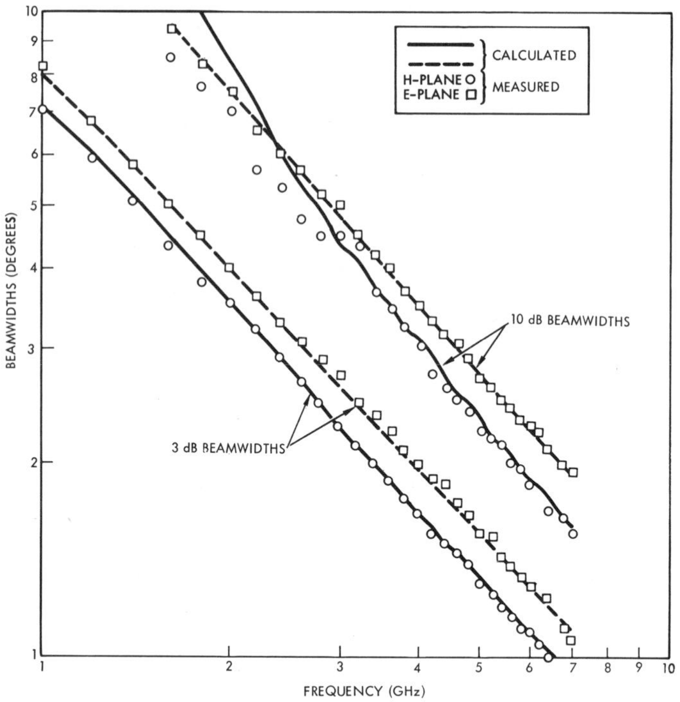

Fig. 2.31. Measured and calculated 3 and 10 dB E- and H-plane beamwidths for antenna No. 5 feeding a 2.74 m $f/D = 0.4$ parabolic reflector

apex 3 cm in front of the focal point. Measurements were made of VSWR, gain and patterns from 1 to 7 GHz. The calculations only utilized the first-order and scalar theories.

The calculated and measured gains of the LPDA are represented in Fig. 2.30. The gain is normally measured at the rear connector of the antenna while (2.51) describes the calculated gain at the feeder tip. Thus the solid line curve in Fig. 2.30 represents these calculations referenced to the tip. To reference the calculations to the rear, the measured loss of the coax cable is included as well as the effects of increased VSWR related to the feeder tip coaxial junction effects in the 4–7 GHz region. The dotted curve is thus the calculated curve referred to the measurement port and compares quite favorably with the measurements. Measured and calculated secondary E- and H-plane 3 dB and 10 dB beamwidths are plotted in Fig. 2.31 and demonstrate the excellent agreement of theory and experiment. In addition, continuous swept gain measurements were made and there were no anomalous frequency-dependent behavior consisting of beam-splitting or high backlobe levels occuring over very narrow bands of frequencies of the type reported in [2.20].

2.3.3. Multiple LPD Feeds

The subject of multiple feeds in a reflector is very complex and no attempt will be made to treat it in detail. Rather, indications of how the previous theory can be applied to the problem, the usefulness of the LPDA as a multiple feed and a technique for reducing mutual coupling are presented.

The problem of multiple feeds in a parabolic reflector has two parts. One is to design the most efficient single feed, and the second is to reduce the interactions between feeds when used in a multiple-feed configuration.

For a configuration where the beams are spread apart to the point of no or very little beam overlap, each feed performs as if it were the only feed in the dish and there is virtually no feed interactions. However, if beam overlap is desired (for example, cross over at the 3 dB level or higher) then the feeds must be spaced close together and several parameters that affect performance and the ability to physically realize the closer feed spacing must be considered. These parameters are feed size and feed to feed coupling. The LPD feed has an advantage in that it minimizes the physical interference problem. The dipoles can be made to intermesh with adjacent feeds as required by a slight staggering of the dipoles in the axial plane, and the only spacing limitations are the diameter of the LPD feed support stem. Tests have proven the capability of physical feed spacings as small as $\lambda/10$ using LPD structures.

The more constraining design problems occur when operating at frequencies where there are losses due to interaction and coupling between feeds. These losses become progressively worse for feed spacing less than $0.6\,\lambda$.

In order to quantify the losses due to coupling as a function of feed spacing, the loss mechanisms were separated into two categories. The first of these called "stem loss" results when the stems or vertical feed supports of surrounding feeds are excited by the feed of interest. This los can be considerable, particularly at spacings less than $\lambda/3$. The second loss is "dipole coupling loss" which results when dipoles of adjacent feeds are excited by the feed of interest. Only the latter loss is considered in the following.

As an example consider two LPD feeds in a reflector spaced a constant distance apart that are to be operated over a wide frequency band. If contiguous angular coverage is desired, it generally requires that the

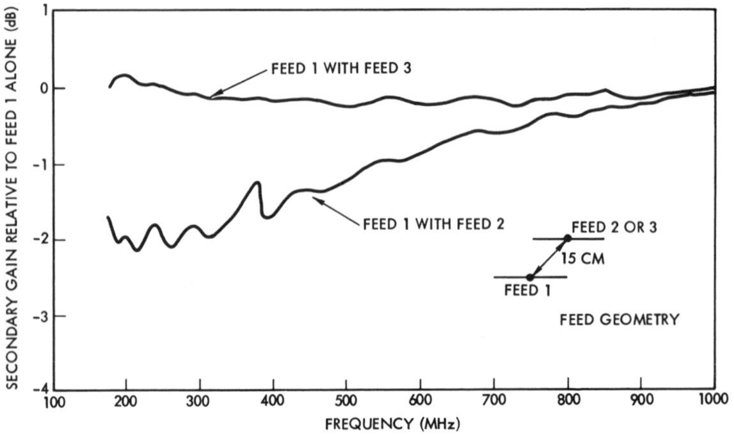

Fig. 2.32. On-axis gain perturbations for coupled LPD antennas as a function of frequency

apertures of the feeds must overlap over a significant portion of their bandwidth so that over the high-frequency portions of the bandwidth, where the secondary beams are narrower, the feeds can be sufficiently close to provide the required coverage. One of the problems then is to prevent or at least reduce the loss on one of the feeds over its entire frequency band, even though part of its aperture is covered by the other feed.

A solution would be to design the active region of one of the antennas to be closer to the focal point of the reflector than the other. The active regions on both arrays are at a constant distance in terms of wavelengths from the virtual apex of each array. Hence, the relative distances and relative defocusing are constant over their bandwidth. As a consequence, when the arrays are intermeshed the longer dipoles of the first antenna block most of the energy coming from or to the second antenna, while the smaller elements of the second antenna do not effectively couple energy away from the first antenna nor disturb its radiation characteristics significantly. When the second antenna is not intermeshed, it suffers only an increased defocusing loss due to its active region being farther back from the focal point.

The concept is confirmed in principle in Fig. 2.32. Antenna No. 2 of Table 2.1 is chosen as the principle feed and is designated as Feed 1 on the figure. It has a $\sigma = 0.06$ and $\tau = 0.825$. Feed 2 has the same antenna parameters only the dipoles are scaled by $\sqrt{\tau}$ so that they are located at the mean distance between corresponding dipoles of the principle feed. Hence, these antennas can intermesh in a co-planar fashion with the maximum distance separating all the dipoles. Feed 3 has a different σ

Table 2.1. Parameter description of antennas used for theoretical and experimental comparison

	Antenna No. 1	Antenna No. 2	Antenna No. 3	Antenna No. 4	Antenna No. 5
Z_0	50 Ω	200 Ω	158 Ω	Tapered line	120 Ω
τ	0.93	0.825	0.87	0.87	0.87055
α	5.7°	36.1°	15.2°	10.7°	14.9°
h_{max}	444.3 cm	86.7	17.2 cm	5.08 cm	14.2 cm
N	12	14	12	8	20
h/a	150	75	109	71.4	60
TERM	50 Ω	Short	Short	Short	100 Ω

Z_0 = unloaded feed line characteristic impedance.
τ = scaling factor that relates the ratio of the length of a dipole or interconnecting transmission line and the next larger dipole or feeder line section.
α = half-angles of the conical envelope of the tips of the feed elements.
h_{max} = half-length of longest dipole in cm.
N = numbers of dipoles in LPD.
h/a = dipole half-length to radius ratio.
TERM = termination—either a resistive termination at the longest dipole or a short 1/8 wavelength behind longest dipole at its resonant frequency.

showing what is termed three notches of separation. That is, a notch is considered as changing the σ by $(\tau)^{-1}$ which is the same as taking Feed 2 and shortening the elements by τ. Feed 3 has $\sigma = 0.1068$. The calculations were made with a $f/D = 0.4$ reflector and the apex of both LPD feeds at the focal point.

Calculations could be made in either of two ways. One, treating the composite pattern from the two feeds as the primary pattern for the reflector and making use of (2.47) and a two-dimensional integration of (2.46) or computing currents on each LPD with the other present but computing the far field of each antenna alone and using (2.51) to compute the secondary fields. The secondary fields are then coherently summed to obtain the composit radiation patterns. The latter method was chosen with the first-order approximation for the solution of the LPD's for the example in Fig. 2.32, and has been proved to be quite adequate for most engineering applications.

References

2.1. R. L. Carrell: "Analysis and Design of the Log-Periodic Dipole Antenna"; Ph. D. Dissertation, University of Illinois, Urbana, Ill. (1960).
2.2. R. F. Harrington: *Field Computation by Moment Methods* (The Macmillan Company, New York, 1968).
2.3. R. F. Harrington, J. R. Mautz: IEEE Trans. Antennas Propagation AP-**15**, 502 (1967).

2.4. J. H. RICHMOND: "Computer Analysis of Three-Dimensional Wire Antennas"; Tech. Report 2708-4, Ohio State University, Columbus, Ohio (1969).
2.5. L. KANTOROVICH, V. KRYLOV: *Approximate Methods of Higher Analysis* (J. Wiley and Sons, New York, 1964), Chapter IV.
2.6. V. H. RUMSEY: Phys. Rev. **94**, 1483 (1954).
2.7. D. S. JONES: IRE Trans. Antennas Propagation AP-**4**, 297 (1956).
2.8. E. C. JORDAN: *Electromagnetic Waves and Radiating Systems* (Prentice-Hall, New York, 1950).
2.9. H. E. KING: IRE Trans. Antennas Propagation AP-**5**, 306 (1957).
2.10. W. A. IMBRIALE, P. G. INGERSON: IEEE Trans. Antennas Propagation AP-**21**, 363 (1973).
2.11. W. M. CHEONG, R. W. KING: Radio Sci. **2**, 1315 (1967).
2.12. W. M. CHEONG, R. W. KING: Radio Sci. **2**, 1303 (1967).
2.13. R. H. KYLE: IEEE Trans. Antennas Propagation AP-**18**, 15 (1970).
2.14. W. A. IMBRIALE: IEEE Trans. Antennas Propagation AP-**23** (1975), to be published.
2.15. S. UDA: J. IEEE (Japan) **452**, 273 (1926).
2.16. H. YAGI: Proc. IRE **16**, 715 (1928).
2.17. G. A. THIELE: IEEE Trans. Antennas Propagation AP-**17**, 24 (1969).
2.18. W. A. IMBRIALE, P. G. INGERSON, W. C. WONG: IEEE Trans. Antennas Propagation AP-**22**, 742 (1974).
2.19. J. RUSE: IEEE Trans. Antennas Propagation AP-**13**, 660 (1965).
2.20. C. C. BANTIN, K. G. BALMAIN: IEEE Trans. Antennas Propagation AP-**18**, 195 (1970).
2.21. K. G. BALMAIN, C. C. BANTIN, C. R. OAKES, L. DAVID: IEEE Trans. Antennas Propagation AP-**19**, 286 (1971).
2.22. W. A. IMBRIALE, P. G. INGERSON: "Optimum Design of Broadband Parabolic Reflector Antennas Fed with Log-Periodic Dipole Arrays"; (Abstracts 23rd Annual Symposium, USAF Antenna Research and Development Program, 1973).

3. Characteristic Modes for Antennas and Scatterers

R. F. HARRINGTON

With 10 Figures

Characteristic modes have long been used for the analysis of electromagnetic problems involving conducting bodies whose surfaces coincide with coordinate surfaces in which the Helmholtz equation is separable. GARBACZ [3.1, 2] has shown that similar modes can be defined for material bodies of arbitrary shape. He approached the problem by diagonalizing the scattering matrix. The same modes were obtained by HARRINGTON and MAUTZ [3.3] by diagonalizing the generalized impedance matrix of the body. This approach provides a more direct derivation of the theory, and is the one taken in this chapter.

The characteristic currents are defined to be the eigenfunctions of a particular weighted eigenvalue equation. For conducting bodies they have the following important properties: a) The characteristic currents are real, or equiphasal, over the surface on which they exist, b) they form an orthogonal set over this surface, and c) they diagonalize the generalized impedance matrix for this surface. The characteristic fields are defined to be the electromagnetic fields produced by the characteristic currents. They have the following important properties: a) The characteristic electric fields have equiphasal tangential components over the surface of the body, b) they form an orthogonal set over the radiation sphere, and c) they diagonalize the scattering matrix for the body. Characteristic modes for dielectric and magnetic bodies can be defined in an analogous manner, and have properties similar to those for conducting bodies.

Because of the above-mentioned special properties, characteristic modes are particularly useful in problems of analysis, synthesis, and optimization of antennas and scatterers. They lead to modal solutions which have the following properties: a) Matrix inversion is not required for computation of the current, since the generalized impedance matrix is diagonalized. b) Pattern synthesis can be accomplished without matrix inversion, since the scattering matrix is diagonalized. c) The current on the body can be controlled using the method of modal resonance. It is shown in Section 3.5 that any desired real current can be made the dominant mode of the body.

3.1. Characteristic Modes for Conducting Bodies

Consider the problem of one or more conducting bodies defined by the surface S in an impressed electric field $\boldsymbol{E}^{\mathrm{i}}$. An operator equation for the current $\boldsymbol{J}$ on S is

$$[L(\boldsymbol{J}) - \boldsymbol{E}^{\mathrm{i}}]_{\tan} = 0 \tag{3.1}$$

where the subscript "tan" denotes the tangential components on S [3.4]. The operator L is defined by

$$L(\boldsymbol{J}) = j\omega \boldsymbol{A}(\boldsymbol{J}) + \nabla \phi(\boldsymbol{J}), \tag{3.2}$$

$$\boldsymbol{A}(\boldsymbol{J}) = \mu \oiint_S \boldsymbol{J}(\boldsymbol{r}') \psi(\boldsymbol{r}, \boldsymbol{r}') \, ds', \tag{3.3}$$

$$\phi(\boldsymbol{J}) = -(j\omega\varepsilon)^{-1} \oiint \nabla' \cdot \boldsymbol{J}(\boldsymbol{r}') \psi(\boldsymbol{r}, \boldsymbol{r}') \, ds', \tag{3.4}$$

$$\psi(\boldsymbol{r}, \boldsymbol{r}') = (4\pi|\boldsymbol{r} - \boldsymbol{r}'|)^{-1} \exp(-jk|\boldsymbol{r} - \boldsymbol{r}'|). \tag{3.5}$$

Here $\boldsymbol{r}$ denotes a field point, $\boldsymbol{r}'$ a source point, and ε, μ, k the permittivity, permeability, and wavenumber, respectively, of free space. Physically, $-L(\boldsymbol{J})$ gives the electric intensity $\boldsymbol{E}$ at any point in space due to the current $\boldsymbol{J}$ on S. In an antenna problem the impressed field $\boldsymbol{E}^{\mathrm{i}}$ is the negative of the tangential component of $\boldsymbol{E}$ over S, assumed known. In a scattering problem the impressed field $\boldsymbol{E}^{\mathrm{i}}$ is due to known sources external to S.

The symmetric product of two vector functions $\boldsymbol{B}$ and $\boldsymbol{C}$ on S is defined as

$$\langle \boldsymbol{B}, \boldsymbol{C} \rangle = \oiint_S \boldsymbol{B} \cdot \boldsymbol{C} \, ds. \tag{3.6}$$

The product $\langle \boldsymbol{B}^*, \boldsymbol{C} \rangle$, where * denotes complex conjugate, defines an inner product for the complex Hilbert space of all square integrable vector functions on S. The operator appearing in (3.1) has the dimensions of impedance, and the operator Z is defined as

$$Z(\boldsymbol{J}) = [L(\boldsymbol{J})]_{\tan}. \tag{3.7}$$

That Z is a symmetric operator, i.e., $\langle \boldsymbol{B}, Z\boldsymbol{C} \rangle = \langle Z\boldsymbol{B}, \boldsymbol{C} \rangle$, follows from the reciprocity theorem [3.5]. However, Z is not an Hermitian operator, i.e., $\langle \boldsymbol{B}^*, Z\boldsymbol{C} \rangle \neq \langle Z^*\boldsymbol{B}^*, \boldsymbol{C} \rangle$. Because Z is symmetric, its Hermitian parts are real and given by

$$R = (Z + Z^*)/2\,, \tag{3.8}$$

$$X = (Z - Z^*)/2j\,. \tag{3.9}$$

Now $Z = R + jX$, where R and X are real symmetric operators. Furthermore, R is positive semi-definite since the power radiated by a current $\boldsymbol{J}$ on S is $\langle \boldsymbol{J}^*, R\boldsymbol{J}\rangle \geqq 0$. If no resonator fields exist internal to S, then R is positive definite, i.e., all currents radiate some power, however small.

3.1.1. Characteristic Currents

Now consider the eigenvalue equation

$$Z(\boldsymbol{J}_n) = v_n R(\boldsymbol{J}_n) \tag{3.10}$$

where v_n are eigenvalues, $\boldsymbol{J}_n$ are eigenfunctions, and R serves as a weight operator. The reason for this particular choice of weight operator is to obtain orthogonality of the radiation fields over the sphere at infinity. Next, let $Z = R + jX$ and $v_n = 1 + j\lambda_n$ in (3.10), and cancel the common term $R(\boldsymbol{J}_n)$. The result is

$$X(\boldsymbol{J}_n) = \lambda_n R(\boldsymbol{J}_n)\,. \tag{3.11}$$

This is the basic eigenvalue equation defining the characteristic currents $\boldsymbol{J}_n$. Since X and R are real symmetric operators, all eigenvalues λ_n of (3.11) must be real, and all eigenfunctions $\boldsymbol{J}_n$ can be chosen real. Then (3.10) states that the tangential electric field on S (left-hand side) is equiphase, since the right-hand side is $1 + j\lambda_n$ times a real quantity.

The characteristic currents $\boldsymbol{J}_n$ obey the usual orthogonality conditions, and furthermore can be normalized such that $\langle \boldsymbol{J}_n^*, R\boldsymbol{J}_n\rangle = 1$. Then the orthogonality conditions can be summarized as

$$\langle \boldsymbol{J}_m^*, R\boldsymbol{J}_n\rangle = \delta_{mn}\,, \tag{3.12}$$

$$\langle \boldsymbol{J}_m^*, X\boldsymbol{J}_n\rangle = \lambda_n \delta_{mn}\,, \tag{3.13}$$

$$\langle \boldsymbol{J}_m^*, Z\boldsymbol{J}_n\rangle = (1 + j\lambda_n)\,\delta_{mn}\,, \tag{3.14}$$

where δ_{mn} is the Kronecker delta (0 if $m \neq n$, and 1 if $m = n$). If the eigencurrents are taken to be real, the conjugate operation on the first terms of (3.12) to (3.14) can be dropped. Henceforth, the characteristic currents will be assumed real.

3.1.2. Characteristic Fields

The electric field $\boldsymbol{E}_n$ and the magnetic field $\boldsymbol{H}_n$ produced by a characteristic current $\boldsymbol{J}_n$ on S are called the characteristic fields. The set of all $\boldsymbol{E}_n$ and $\boldsymbol{H}_n$ form a Hilbert space of all fields produced by currents on S. Orthogonality relationships for the characteristic fields can be obtained from those for characteristic currents by means of the complex Poynting theorem [3.5]. The complex power balance for currents $\boldsymbol{J}$ on S is given by

$$\begin{aligned} P &= \langle \boldsymbol{J}^*, Z\boldsymbol{J}\rangle = \langle \boldsymbol{J}^*, R\boldsymbol{J}\rangle + j\langle \boldsymbol{J}^*, X\boldsymbol{J}\rangle \\ &= \oiint_{S'} \boldsymbol{E}\times\boldsymbol{H}^*\cdot d\boldsymbol{s} + j\omega \iiint_{\tau'} (\mu \boldsymbol{H}\cdot\boldsymbol{H}^* - \varepsilon \boldsymbol{E}\cdot\boldsymbol{E}^*)\, d\tau\,, \end{aligned} \tag{3.15}$$

where S' is any surface enclosing S, and τ' is the region enclosed by S'. Equation (3.15) is a Hermitian quadratic form for which the associated Hermitian bilinear form is

$$P(\boldsymbol{J}_m, \boldsymbol{J}_n) = \langle \boldsymbol{J}_m^*.\ Z\boldsymbol{J}_n\rangle\,. \tag{3.16}$$

If $\boldsymbol{J}_m$ and $\boldsymbol{J}_n$ are eigencurrents, then the orthogonality relationships (3.12) to (3.14) apply and (3.16) becomes

$$\begin{aligned} &\oiint_{S'} \boldsymbol{E}_m\times\boldsymbol{H}_n^*\cdot d\boldsymbol{s} + j\omega \iiint_{\tau'} (\mu \boldsymbol{H}_m\cdot\boldsymbol{H}_n^* - \varepsilon \boldsymbol{E}_m\cdot\boldsymbol{E}_n^*)\, d\tau \\ &= (1 + j\lambda_n)\, \delta_{mn}\,. \end{aligned} \tag{3.17}$$

This can be separated into real and imaginary parts to give orthogonality relationships analogous to (3.12) and (3.13).

If the surface S is of finite extent and if S' is chosen to be the radiation sphere S_∞, then the real part of (3.17) leads to an orthogonality of radiation fields. On S_∞, both $\boldsymbol{E}$ and $\boldsymbol{H}$ are of the form of outward traveling waves, and $\boldsymbol{E}_n = (\mu/\varepsilon)^{1/2}\, \boldsymbol{H}_n \times \hat{\boldsymbol{n}}$, where $\hat{\boldsymbol{n}}$ is the unit radial vector. Addition of (3.17) to its conjugate with m and n interchanged yields

$$(\varepsilon/\mu)^{1/2} \oiint_{S_\infty} \boldsymbol{E}_m\cdot\boldsymbol{E}_n^*\, ds = \delta_{mn}\,. \tag{3.18}$$

Hence, the characteristic electric fields form an orthonormal set in the Hilbert space of all square-integrable vector functions on S_∞. Similarly, (3.18) can be expressed in terms of characteristic magnetic fields as

$$(\mu/\varepsilon)^{1/2} \oiint \boldsymbol{H}_m\cdot\boldsymbol{H}_n^*\, ds = \delta_{mn}\,. \tag{3.19}$$

Finally, an orthogonality relation of mutual energy is obtained by subtracting (3.17) from its conjugate with m and n interchanged. This is

$$\omega \iiint (\mu \boldsymbol{H}_m \cdot \boldsymbol{H}_n^* - \varepsilon \boldsymbol{E}_m \cdot \boldsymbol{E}_n^*)\, d\tau = \lambda_n \delta_{mn}, \tag{3.20}$$

where the integration extends over all space. For $m = n$, (3.20) states that λ_n is 2ω times the total stored magnetic energy minus the total stored electric energy.

3.2. Modal Solutions

A modal solution for the current $\boldsymbol{J}$ on the conducting body is obtained by using the characteristic currents as both expansion and testing functions in the method of moments [3.4]. Hence, $\boldsymbol{J}$ is assumed to be a linear superposition of the mode currents

$$\boldsymbol{J} = \sum_n \alpha_n \boldsymbol{J}_n, \tag{3.21}$$

where the α_n are coefficients to be determined. Equation (3.21) is substituted into (3.1) to obtain

$$\left[\sum_n \alpha_n L(\boldsymbol{J}_n) - \boldsymbol{E}^{\mathrm{i}}\right]_{\tan} = 0. \tag{3.22}$$

Next, the inner product of (3.22) with each $\boldsymbol{J}_m$ in turn is taken, giving the set of equations

$$\sum_n \alpha_n \langle \boldsymbol{J}_m, Z\boldsymbol{J}_n\rangle - \langle \boldsymbol{J}_m, \boldsymbol{E}^{\mathrm{i}}\rangle = 0, \tag{3.23}$$

where $m = 1, 2, \ldots$. Here the subscript "tan" has been dropped, and $L_{\tan} = Z$ has been used. Because of the orthonormality relationship (3.14), Eq. (3.23) reduces to

$$\alpha_n(1 + j\lambda_n) = \langle \boldsymbol{J}_n, \boldsymbol{E}^{\mathrm{i}}\rangle. \tag{3.24}$$

The right-hand side of (3.24) is called the modal excitation coefficient

$$V_n^{\mathrm{i}} = \langle \boldsymbol{J}_n, \boldsymbol{E}^{\mathrm{i}}\rangle = \oiint_S \boldsymbol{J}_n \cdot \boldsymbol{E}^{\mathrm{i}}\, ds. \tag{3.25}$$

Now if α_n from (3.23) is substituted into (3.21), one has the modal solution for $\boldsymbol{J}$ on S

$$\boldsymbol{J} = \sum_n V_n^{\mathrm{i}} \boldsymbol{J}_n (1 + j\lambda_n)^{-1}. \tag{3.26}$$

The fields are linearly related to the currents, and hence also can be expressed in modal form. Explicitly, these forms are

$$E = \sum_n V_n^{\mathrm{i}} E_n (1 + j\lambda_n)^{-1}, \tag{3.27}$$

$$H = \sum_n V_n^{\mathrm{i}} H_n (1 + j\lambda_n)^{-1}, \tag{3.28}$$

where E and H are the fields from J. If the eigencurrents are not normalized, the terms $(1 + j\lambda_n)$ in (3.26) through (3.28) must be replaced by $(1 + j\lambda_n)\langle J_n, RJ_n\rangle$.

If the reciprocity theorem is used, alternative expressions for the modal excitation coefficients are obtained. For example, if E^{i} is produced by an impressed electric current J^{i}, then reciprocal to (3.25) one has

$$V_n^{\mathrm{i}} = \iiint E_n \cdot J^{\mathrm{i}}\, d\tau, \tag{3.29}$$

where the integration extends over the impressed currents. Similarly, if E^{i} is produced by an impressed magnetic current M^{i}, then reciprocal to (3.25) one has

$$V_n^{\mathrm{i}} = -\iiint H_n \cdot M^{\mathrm{i}}\, d\tau. \tag{3.30}$$

More generally, if E^{i} is produced by both electric currents J^{i} and magnetic currents M^{i}, then V_n^{i} is given by the sum of the right-hand sides of (3.29) and (3.30).

3.2.1. Linear Measurements

Any scalar ϱ linearly related to the current, i.e., a linear functional of the current, will be called a linear measurement of the current. Two examples of linear measurements are a) a component of the current at some point on S, and b) a component of the field (E or H) at some point in space. Every linear functional of J can be expressed as

$$\varrho = \langle E^{\mathrm{m}}, J\rangle, \tag{3.31}$$

where E^{m} is a given vector function, usually an electric field on S. For example, if ϱ is the jth component of the field E_j^J from J, then (3.31) becomes [3.4, 5]

$$E_j^J = \langle E^j, J\rangle, \tag{3.32}$$

where $\boldsymbol{E}^j$ is the electric field on S produced by a j-directed electric dipole $Il = 1$ placed at the field point. If the jth component of $\boldsymbol{H}$ were desired, then a unit magnetic dipole would be placed at the field point, and so on.

If the modal solution (3.26) is substituted into the linear measurement formula (3.31), there results

$$\varrho = \sum_n V_n^{\mathrm{i}} V_n^{\mathrm{m}} (1 + j\lambda_n)^{-1}, \tag{3.33}$$

where V_n^{m} is the modal measurement coefficient

$$V_n^{\mathrm{m}} = \langle \boldsymbol{J}_n, \boldsymbol{E}^{\mathrm{m}} \rangle = \oiint \boldsymbol{J}_n \cdot \boldsymbol{E}^{\mathrm{m}} \, ds \, . \tag{3.34}$$

Note that V_n^{m} is of the same functional form as the excitation coefficient V_n^{i} given by (3.25). Hence, (3.33) is a symmetric bilinear functional of $\boldsymbol{E}^{\mathrm{i}}$ and of $\boldsymbol{E}^{\mathrm{m}}$. This symmetry is a consequence of the symmetry of the original operator Z.

Reciprocal forms for the measurement coefficients, analogous to (3.29) and (3.30) for excitation coefficients, can also be written. For example, if the source of $\boldsymbol{E}^{\mathrm{m}}$ is electric current $\boldsymbol{J}^{\mathrm{m}}$, then

$$V_n^{\mathrm{m}} = \iiint \boldsymbol{E}_n \cdot \boldsymbol{J}^{\mathrm{m}} \, d\tau \tag{3.35}$$

analogous to (3.29). If the source of $\boldsymbol{E}^{\mathrm{m}}$ is the magnetic current $\boldsymbol{M}^{\mathrm{m}}$, then

$$V_n^{\mathrm{m}} = - \iiint \boldsymbol{H}_n \cdot \boldsymbol{M}^{\mathrm{m}} \, d\tau \tag{3.36}$$

analogous to (3.30). Finally, if $\boldsymbol{E}^{\mathrm{m}}$ is produced by both a $\boldsymbol{J}^{\mathrm{m}}$ and an $\boldsymbol{M}^{\mathrm{m}}$, the measurement coefficient V_n^{m} is given by the sum of the right-hand sides of (3.35) and (3.36).

3.2.2. Application to Radiation Problems

An important specialization of the general theory is the problem of radiation from apertures in conducting bodies. Consider a conducting body S in which one or more apertures exist. There are sources internal to S which produce a tangential electric field $\boldsymbol{E}_{\tan}$ (assumed known) over the apertures. Then $\boldsymbol{E}^{\mathrm{i}} = -\boldsymbol{E}_{\tan}$ is the impressed field, and the modal excitation coefficients (3.25) become

$$V_n^{\mathrm{i}} = - \oiint \boldsymbol{J}_n \cdot \boldsymbol{E}_{\tan} \, ds \, . \tag{3.37}$$

The radiated field is given by the modal solution (3.27). For computation, one must consider one number at a time, say some component of $\boldsymbol{E}$

at a point $\boldsymbol{r}$. To accomplish this, place a unit electric dipole $I\boldsymbol{l} = \boldsymbol{u}_m$ at $\boldsymbol{r}$ and evaluate the modal measurement coefficient by (3.35). This gives

$$V_n^{\mathrm{m}} = \boldsymbol{E}_n^{\mathrm{m}} \cdot \boldsymbol{u}_m , \tag{3.38}$$

where $\boldsymbol{E}^{\mathrm{m}}$ is the field produced by the dipole. If the dipole lies on the radiation sphere, then in the vicinity of S its field is [3.5]

$$\boldsymbol{E}^{\mathrm{m}} = -j\omega\mu(4\pi r_m)^{-1} \exp(-jkr_m)\,[\boldsymbol{u}_m \exp(-j\boldsymbol{k}_m \cdot \boldsymbol{r})] . \tag{3.39}$$

Here $\boldsymbol{k}_m$ is the vector propagation constant of the wave from the distant dipole, and $\boldsymbol{r}_m$ is the position vector to it. Now (3.38) becomes

$$V_n^{\mathrm{m}} = -j\omega\mu(4\pi r_m)^{-1} \exp(-jkr_m) \oiint \boldsymbol{J}_n \cdot \boldsymbol{u}_m \exp(-j\boldsymbol{k}_m \cdot \boldsymbol{r})\, ds \tag{3.40}$$

This is next substituted into (3.27) and the result dotted into $\boldsymbol{u}_m$ to give

$$\boldsymbol{E} \cdot \boldsymbol{u}_m = -j\omega\mu(4\pi r_m)^{-1} \exp(-jkr_m) \sum_n V_n^{\mathrm{i}} R_n^{\mathrm{m}} (1 + j\lambda_n)^{-1} , \tag{3.41}$$

where the V_n^{i} are given by (3.37), and

$$R_n^{\mathrm{m}} = \iint \boldsymbol{J}_n \cdot \boldsymbol{u}_m \exp(-j\boldsymbol{k}_m \cdot \boldsymbol{r})\, ds \tag{3.42}$$

are called the plane-wave measurement coefficients. Equation (3.41) provides a formula convenient for computation.

3.2.3. Application to Scattering Problems

Another important specialization of the general solution is the problem of plane-wave scattering by conducting bodies. Consider a conducting body S in a plane-wave field, for which the impressed field is the unit plane wave

$$\boldsymbol{E}^{\mathrm{i}} = \boldsymbol{u}_i \exp(-j\boldsymbol{k}_i \cdot \boldsymbol{r}) . \tag{3.43}$$

Here $\boldsymbol{u}_i$ is the polarization vector, and $\boldsymbol{k}_i$ is the propagation vector. The excitation coefficients (3.25) are now

$$V_n^{\mathrm{i}} = R_n^{\mathrm{i}} = \oiint_S \boldsymbol{J}_n \cdot \boldsymbol{u}_i \exp(-j\boldsymbol{k}_i \cdot \boldsymbol{r})\, ds . \tag{3.44}$$

Note that this is the same functional form as the plane-wave measurement coefficients (3.42). The determination of the scattered field at some measurement point $\boldsymbol{r}_m$ is the same problem as the determination of the radiation field in the antenna problem. Hence, the field scattered in the direction $\boldsymbol{r}_m$ is given by (3.41) with V_n^{i} replaced by R_n^{i}, or

$$\boldsymbol{E}\cdot\boldsymbol{u}_m = -j\omega\mu(4\pi r_m)^{-1}\exp(-jkr_m)\sum_n R_n^{\mathrm{i}}\,R_n^{\mathrm{m}}(1+j\lambda_n)^{-1}\,. \tag{3.45}$$

A commonly used parameter in plane-wave scattering problems is the echo area, defined as [3.5]

$$\sigma = 4\pi r_m^2\,|\boldsymbol{E}\cdot\boldsymbol{u}_m|^2\,. \tag{3.46}$$

A substitution from (3.45) into (3.46) gives

$$\sigma = (\omega\mu)^2\,(4\pi)^{-1}\left|\sum_n R_n^{\mathrm{i}}\,R_n^{m}(1+j\lambda_n)^{-1}\right|^2. \tag{3.47}$$

Note that σ is a function of the polarization of the incident wave and of the measurement wave, as well as the incident wave direction and the measurement wave direction.

3.3. Scattering Matrices

The scattering matrix was first defined as that matrix which relates the amplitudes of incoming spherical modes to outgoing spherical modes [3.6]. More generally, the incoming and outgoing waves can be expanded in terms of arbitrary basis functions. It is shown below that if the characteristic fields $\boldsymbol{E}_n$ are chosen as the basis for outgoing waves, and their conjugates $\boldsymbol{E}_n^*$ as the basis for incoming waves, then the scattering matrix is diagonalized.

In a scattering problem the far-zone field can be expressed as the sum of incoming and outgoing waves as

$$\boldsymbol{E} = \boldsymbol{E}_{\mathrm{in}} + \boldsymbol{E}_{\mathrm{out}}\,. \tag{3.48}$$

For a given scatterer, for each incoming wave $\boldsymbol{E}_{\mathrm{in}}$ there is a unique outgoing wave $\boldsymbol{E}_{\mathrm{out}}$. The scattering operator is that which operates on $\boldsymbol{E}_{\mathrm{in}}$ to give $\boldsymbol{E}_{\mathrm{out}}$, i.e.,

$$\boldsymbol{E}_{\mathrm{out}} = S\boldsymbol{E}_{\mathrm{in}}\,. \tag{3.49}$$

Given an outgoing wave, the conjugate of it will be an incoming wave. The characteristic fields $\boldsymbol{E}_n$ are outgoing waves, and we choose them as basis functions for $\boldsymbol{E}_{\text{out}}$, i.e.,

$$\boldsymbol{E}_{\text{out}} = \sum_n b_n \boldsymbol{E}_n . \tag{3.50}$$

Their conjugates $\boldsymbol{E}_n^*$ are incoming waves, and we choose them as basis functions for $\boldsymbol{E}_{\text{in}}$, i.e.,

$$\boldsymbol{E}_{\text{in}} = \sum_n a_n \boldsymbol{E}_n^* . \tag{3.51}$$

The scattering matrix $[S]$ is that which relates the column vector $\bar{\boldsymbol{b}}$ (components b_n) to the column vector $\bar{\boldsymbol{a}}$ (components a_n) according to

$$\bar{\boldsymbol{b}} = [S]\, \bar{\boldsymbol{a}} . \tag{3.52}$$

When no body is present, $\bar{\boldsymbol{b}} = \bar{\boldsymbol{a}}$ and the scattering matrix is the identity matrix. This is evident by noting that $\boldsymbol{E}_n + \boldsymbol{E}_n^*$ is a source-free field, sometimes called a standing wave field.

We next show that for perfectly conducting bodies $[S]$ is a diagonal matrix, and obtain its elements. The impressed field $\boldsymbol{E}^{\text{i}}$ is a free-space field, and hence must be of the form

$$\boldsymbol{E}^{\text{i}} = \sum_n a_n (\boldsymbol{E}_n + \boldsymbol{E}_n^*) . \tag{3.53}$$

The incoming wave is given by (3.51). The outgoing wave is given by the sum of the scattered field (3.27) plus the outgoing part of (3.53). Hence, the coefficients of (3.50) are

$$b_n = a_n + V_n^{\text{i}} (1 + j\lambda_n)^{-1} . \tag{3.54}$$

The modal excitation coefficients are next found from (3.25) and (3.53) as

$$V_n^{\text{i}} = -2a_n . \tag{3.55}$$

Substituting (3.55) into (3.54) and simplifying, one has

$$b_n = -a_n (1 - j\lambda_n)(1 + j\lambda_n)^{-1} . \tag{3.56}$$

Therefore $[S]$, as defined by (3.52), is a diagonal matrix with diagonal elements equal to $-(1 - j\lambda_n)(1 + j\lambda_n)^{-1}$.

3.4. Computation of Characteristic Modes

For computation, the operator equations must be reduced to matrix equations, which can be conveniently done by the method of moments [3.4]. This section is a condensation of the more complete exposition available in the literature [3.7].

The mode currents are real, and hence a set of real expansion functions $\boldsymbol{W}_j$ are used for the $\boldsymbol{J}_n$, i.e.,

$$\boldsymbol{J}_n = \sum_j I_j \boldsymbol{W}_j \,, \tag{3.57}$$

where the I_j are real coefficients to be determined. Substitute (3.57) into (3.11) and use the linearity of the operators to obtain

$$\sum_j I_j X \boldsymbol{W}_j = \lambda_n \sum_j I_j R \boldsymbol{W}_j \,. \tag{3.58}$$

To obtain symmetric matrices, the same $\boldsymbol{W}_i$ are used as testing functions. Taking the inner product of (3.58) with each $\boldsymbol{W}_i$ in turn, one has the set of equations

$$\sum_j I_j \langle \boldsymbol{W}_i, X \boldsymbol{W}_j \rangle = \lambda_n \sum_j I_j \langle \boldsymbol{W}_i, R \boldsymbol{W}_j \rangle \,, \tag{3.59}$$

where $i = 1, 2, \ldots$. These equations can be written as the matrix eigenvalue equation

$$[X]\,\bar{\boldsymbol{I}}_n = \lambda_n [R]\,\bar{\boldsymbol{I}}_n \,, \tag{3.60}$$

where $\bar{\boldsymbol{I}}_n$ is a column vector of the I_i, $[X]$ is a square matrix with elements $\langle \boldsymbol{W}_i, X \boldsymbol{W}_j \rangle$, and $[R]$ is a square matrix with elements $\langle \boldsymbol{W}_i, R \boldsymbol{W}_j \rangle$. Equation (3.60) is a real symmetric weighted matrix eigenvalue equation. Its eigenvalues λ_n approximate those of the operator Eq. (3.11), and its eigenvectors $\bar{\boldsymbol{I}}_n$ define functions according to (3.57) which approximate the eigenfunctions of (3.11).

The corresponding matrix approximation to the complex eigenvalue equation (3.10) is

$$[Z]\,\bar{\boldsymbol{I}}_n = (1 + j\lambda_n)\,[R]\,\bar{\boldsymbol{I}}_n \,, \tag{3.61}$$

where $[Z] = [R + jX]$ is the generalized impedance matrix of the body [3.4]. It has been evaluated for a number of cases [3.8, 9]. If the $\boldsymbol{W}_i$ are differentiable, a convenient formula for the impedance elements is

$$Z_{ij} = \oiint_S ds' \oiint_S ds[j\omega\mu \boldsymbol{W}_i' \cdot \boldsymbol{W}_j + (j\omega\varepsilon)^{-1} (\nabla' \cdot \boldsymbol{W}_i')(\nabla \cdot \boldsymbol{W}_j)]\,\psi_Z, \tag{3.62}$$

where the prime denotes functions of the primed coordinates and

$$\psi_Z = (4\pi |\boldsymbol{r} - \boldsymbol{r}'|)^{-1} \exp(-jk|\boldsymbol{r} - \boldsymbol{r}'|). \tag{3.63}$$

The Hermitian parts of $[Z]$ are its real part $[R]$ and its imaginary part $[X]$, obtained in the usual way. In particular, the elements R_{ij} are given by (3.62) with ψ_Z replaced by

$$\psi_R = (j4\pi |\boldsymbol{r} - \boldsymbol{r}'|)^{-1} \sin(k|\boldsymbol{r} - \boldsymbol{r}'|) \tag{3.64}$$

and the elements X_{ij} are given by (3.62) with ψ_Z replaced by

$$\psi_X = (4\pi |\boldsymbol{r} - \boldsymbol{r}'|)^{-1} \cos(k|\boldsymbol{r} - \boldsymbol{r}'|). \tag{3.65}$$

Numerical evaluation of these elements is considered in [3.10, 11].

The matrix equivalents of the orthogonality relationships (3.12) to (3.14) are also of interest. For example, that for R is

$$\begin{aligned}\langle \boldsymbol{J}_m^*, R\boldsymbol{J}_n \rangle &= \Big\langle \sum_i I_{i,m}^* \boldsymbol{W}_i, R \sum_j I_{j,n} \boldsymbol{W}_j \Big\rangle \\ &= \sum_{i,j} I_{i,m}^* I_{j,n} \langle \boldsymbol{W}_i, R\boldsymbol{W}_j \rangle \\ &= \tilde{\boldsymbol{I}}_m^* [R] \boldsymbol{I}_n = \delta_{mn},\end{aligned} \tag{3.66}$$

where the tilde denotes transpose. Similar derivatives hold for X and Z. The resultant matrix orthogonality relationships are

$$\tilde{\boldsymbol{I}}_m^* [R] \boldsymbol{I}_n = \delta_{mn}, \tag{3.67}$$

$$\tilde{\boldsymbol{I}}_m^* [X] \boldsymbol{I}_n = \lambda_n \delta_{mn}, \tag{3.68}$$

$$\tilde{\boldsymbol{I}}_m^* [Z] \boldsymbol{I}_n = (1 + j\lambda_n)\, \delta_{mn}. \tag{3.69}$$

Because the $\boldsymbol{I}_n$ are real, these orthogonality relationships also apply with the conjugation omitted.

The conventional method of computing the eigenvalues λ_n and eigenvectors $\boldsymbol{I}_n$ of (3.60) requires $[R]$ to be positive definite [3.12]. In our problem $[R]$ is positive semi-definite in theory, but because of numerical inaccuracies it may actually be indefinite with some small negative eigenvalues. A modification of the conventional method has been developed and described in detail in [3.7].

3.4.1. Computations for Bodies of Revolution

A general computer program for calculating the characteristic modes of conducting bodies of revolution is available [3.10]. Figure 3.1 defines the coordinate system used for bodies of revolution. The surface S is generated by rotating the contour C about the z axis. The surface coordinates on S are t (length variable along C) and ϕ (angle of rotation from the x axis). The spherical coordinates of a field point are r, θ, ϕ. The current $\boldsymbol{J}$ on S has two components, J_t and J_ϕ. Letting $\boldsymbol{u}_t$ and $\boldsymbol{u}_\phi$ denote unit vectors in the t and ϕ directions, one can choose two sets of real expansion functions

$$\{\boldsymbol{u}_t f_i(t), \boldsymbol{u}_t f_i(t) \cos n\phi, \boldsymbol{u}_\phi f_i(t) \sin n\phi\} \tag{3.70}$$

and

$$\{\boldsymbol{u}_\phi f_i(t), \boldsymbol{u}_t f_i(t) \sin n\phi, -\boldsymbol{u}_\phi f_i(t) \cos n\phi\}, \tag{3.71}$$

where i and n are positive integers. These two sets are sufficiently general to represent an arbitrary $\boldsymbol{J}$ on S if $f_i(t)$ form a complete set in the t domain. If the testing function $\boldsymbol{W}_i$ is from the set (3.70), and the expansion function $\boldsymbol{W}_j$ is from the set (3.71), then the resulting impedance element is zero.

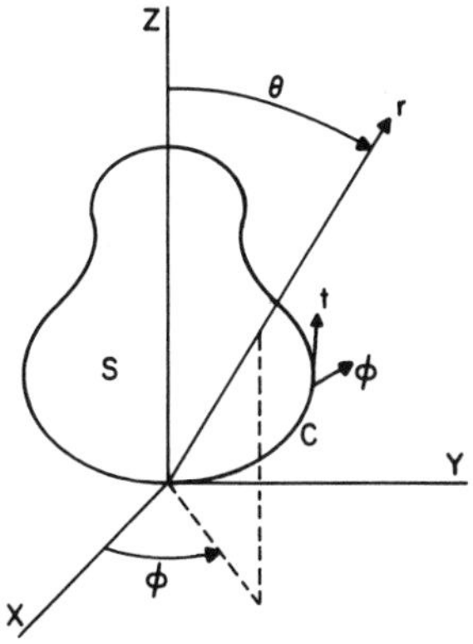

Fig. 3.1. Coordinate system for bodies of revolution

The sets (3.70) and (3.71) can therefore be treated independently. The impedance matrix for each set has a block diagonal form, which further reduces the complexity. Details of the evaluation of the impedance matrices can be obtained from [3.7] and [3.8].

For computation, $2N+1$ nearly equidistant points $t_0, t_1, \ldots, t_{2N}$ are specified along C, with t_0 at the beginning and t_{2N} at the end of C. The $f(t)$ are taken as triangle functions divided by the radius, each extending over four intervals $(t_j - t_{j-1})$. For expansion, the $f_i(t)$ are approximated by four pulses, and for testing they are approximated by four impulses. Because of these approximation, the impedance matrices are not exactly symmetric, as they would be if evaluated exactly. This asymmetry was eliminated by averaging corresponding off-diagonal elements of $[Z]$. A number of computational checks showed that this averaging had no noticeable effect in radiation and scattering problems. To check the computer program, it was first run on conducting spheres, for which the exact modes are known. The numerically evaluated eigenvalues and eigenfunctions agreed with the exact ones to within one percent for the lower-order modes (small $|\lambda_n|$). The accuracy of the numerical procedure decreased for the higher-order modes, as expected.

To illustrate computations for a representative body, consider the cone-sphere formed by a cone of 10° half-angle joined smoothly to a sphere of diameter 0.4 wavelength. The contour is approximated by 40 straight-line segments of equal length, and 19 expansion functions are used for J_t and J_ϕ. Figure 3.2 shows the six lowest order characteristic currents for the rotationally symmetric modes, plotted versus the contour length variable, starting at the tip and ending on the sphere. These modes are the ones which are used in the problem of radiation from rotationally symmetric apertures. The crosses denote t-directed currents and the squares denote ϕ-directed currents. In the rotationally symmetric case, if J_t is nonzero, then $J_\phi = 0$, and vice versa. Note that the first three modes have only J_t, and are similar to modes on a straight wire. The next two modes have only J_ϕ, and are similar to wire loop modes. The value of λ_n in each case is listed under the corresponding graph. The currents are normalized so that their mean-square value over the surface is unity.

Figure 3.3 shows the characteristic gain patterns for the six lowest-order rotationally symmetric modes of the cone-sphere. They are plots of the normalized radiation intensity versus θ from the corresponding mode currents of Fig. 3.2. Tic marks correspond to a gain of 2. The first modes are due to J_t and have only an E_θ in the radiation field. The next two modes are due to J_ϕ and have only an E_ϕ in the radiation field. The three-dimensional pattern is obtained by rotating each plot about the Z axis.

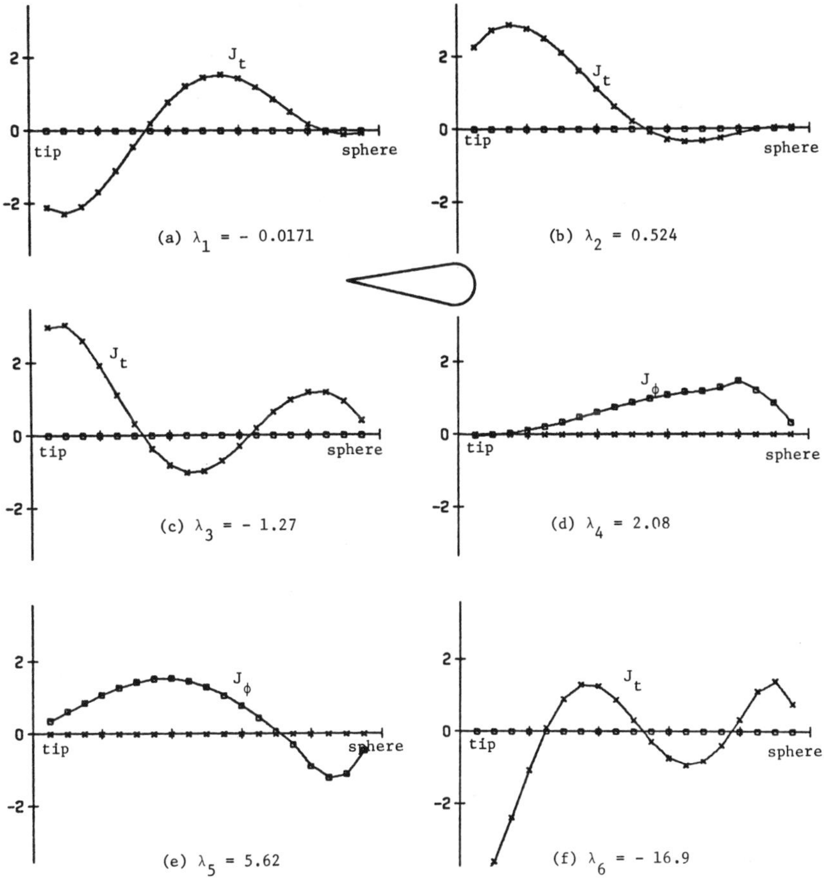

Fig. 3.2a–f. Characteristic currents for a cone-sphere, length 1.36 wavelengths, sphere diameter 0.4 wavelengths. The six lowest-order rotationally symmetric modes are shown

Modes which vary as $\cos\phi$ and $\sin\phi$ are considered in [3.7] for the same cone-sphere. These are the modes that would be used in the problem of scattering due to a plane-wave axially incident on the body. Now each mode has both a J_t and a J_ϕ, in contrast to the rotationally symmetric modes which have only one component. The characteristic mode patterns for these $\cos\phi$, $\sin\phi$ modes also have both an E_θ and E_ϕ component, in contrast to the rotationally symmetric modes which have only one component. Still higher-order modes, varying as $\sin n\phi$ and $\cos n\phi$, are required for more general problems, such as non-axial plane-wave scattering and radiation from arbitrary apertures. These modes also can be computed with the general computer programs [3.10].

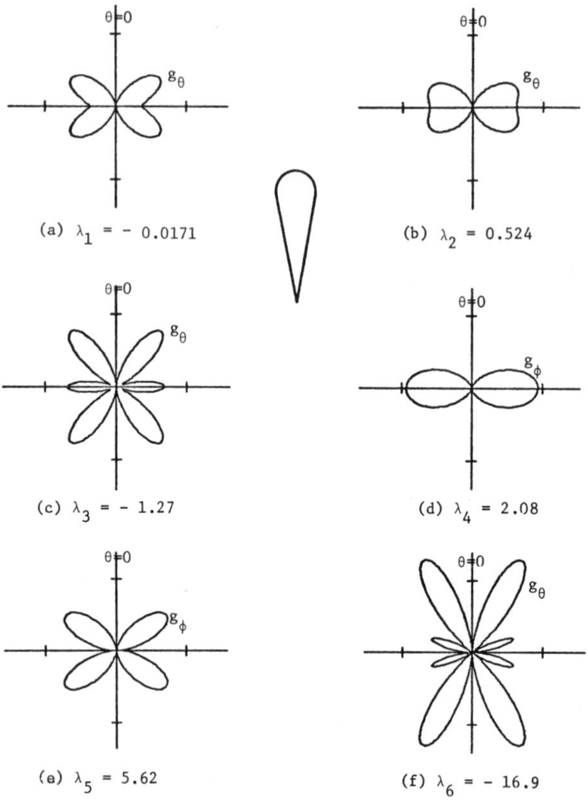

Fig. 3.3a–f. Characteristic gain patterns for a cone-sphere, length 1.36 wavelengths, sphere diameter 0.4 wavelengths. The six lowest-order rotationally symmetric modes are shown

3.4.2. Convergence of Radiation Patterns

It has been implied that only a few modes are needed to characterize the radiation and scattering properties of electrically small and intermediate size bodies. This property has been demonstrated by computing the modal solution for varying numbers of modes, showing that their radiation and scattering patterns converge rapidly to known solutions. Since the characteristic field patterns form an orthogonal set in the Hilbert space of radiation fields on the sphere at infinity, it follows that this convergence is in a least-squares sense.

Figure 3.4 shows the convergence of the modal solution for the cone-sphere excited by a voltage applied across a narrow slot at the cone-to-sphere junction. The solid curve in each figure is the radiation-gain pattern obtained by matrix inversion [3.8], and the crosses are the

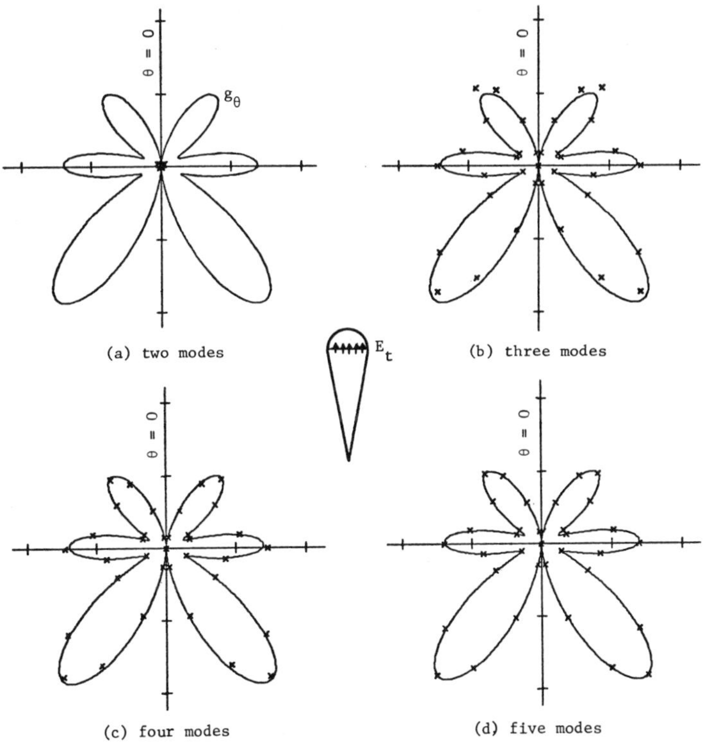

Fig. 3.4a–d. Convergence of the modal solution for the cone-sphere excited by a voltage across a slot at the cone-to-sphere junction. Solid curves are the matrix inversion solution, x's are the modal solutions

corresponding modal solution. The modes used are the rotationally symmetric ones having only J_t on S (Fig. 3.2) and only E_θ in the radiation field (Fig. 3.3). No other modes are excited by this particular slot. The modal solution of Fig. 3.3a uses the two lowest-order modes, and is very small (all crosses lie at the coordinate origin). This could have been predicted from the fact that the two lowest-order characteristic mode currents, Fig. 3.2a and b, have almost zero amplitude at the cone-to-sphere junction, and hence are not excited. The modal solution of Fig. 3.4b uses the three lowest-order modes, and shows that the gain pattern is predominantly that of the third mode. The modal solution of Fig. 3.4c uses the four lowest-order J_t modes, and that of Fig. 3.4d uses the five lowest-order J_t modes. The gain pattern has fully converged when five modes are used.

Convergence of the modal solution for plane-wave scattering by the same cone-sphere was shown in [3.7]. For axial incidence, this solution uses only those modes which vary as $\sin\phi$ and $\cos\phi$. The plane wave excites all such modes, in contrast to aperture problems which often excite only some of the modes. For the particular cone-sphere used, about six modes were needed for convergence of the scattering pattern. If the cone-sphere were excited by a non-axially propagating plane wave, then the rotationally symmetric modes plus those which vary as $\sin n\phi$, $\cos n\phi$ would be excited. However, even in this case, if the body is of electrically small diameter only the rotationally symmetric modes and the $\sin\phi$, $\cos\phi$ modes contribute significantly to the radiation pattern.

3.4.3. Computations for Wire Objects

A general computer program for calculating the characteristic modes of wire objects of arbitrary shape is available [3.11]. The wire in space is specified by a number of points along its axis, plus its diameter. There may be more than one wire present, and these wires may have free ends, may be closed on themselves, or may be joined together. The expansion and testing functions are chosen to be triangle functions extending over four subsections of the wire. For expansion the triangles are approximated by four pulses, and for testing they are approximated by four impulses. Because of these approximations the impedance matrix is not exactly symmetric, as it should be. It is made symmetric by averaging corresponding off-diagonal impedance elements.

Computations of characteristic currents and characteristic fields have been made for a number of wire objects, and these modes have been used in modal solutions to demonstrate convergence in radiation and scattering problems. As an example, consider the wire arrow shown in the central insert of Fig. 3.5. The parameter "a" is 1/4 wavelength, and the wire diameter is 0.004 wavelength. The graphs of Fig. 3.5 show the six lowest-order characteristic currents, plotted as a function of the contour variable, starting at the tip and ending at the tip. Note that the modes are either symmetric (even) or asymmetric (odd) about the midpoint with respect to the wire length variable. This could be predicted from the symmetry of the geometry of the wire. However, this symmetry was not used in the computations because the program is written for objects of arbitrary shape. The currents were normalized by choosing their maximum amplitude to be unity.

Figure 3.6 exhibits the characteristic gain patterns for the six lowest-order characteristic fields of the wire arrow. The arrow is considered to lie in the $x-z$ plane, with its axis along the z axis. The two patterns shown are the gain in the $x=0$ plane (labeled g_x) and the gain in the $y=0$ plane (labeled g_y). The g_y pattern is always θ-polarized, and the g_x

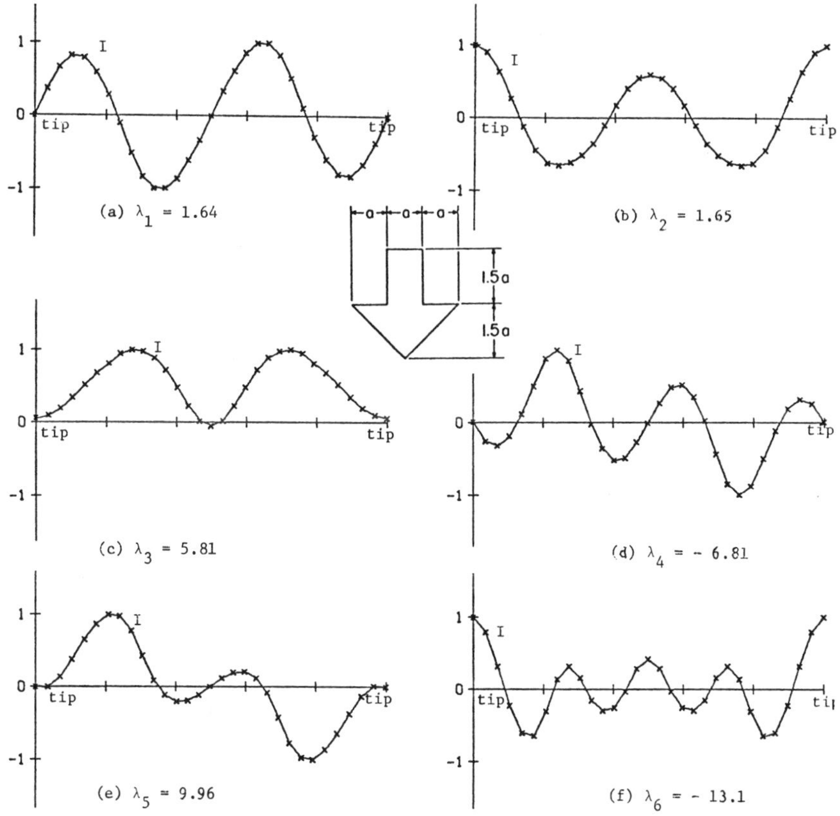

Fig. 3.5a–f. Characteristic currents for the wire arrow, with "a" 0.25 wavelength and wire diameter 0.004 wavelength. The six lowest-order modes are shown

pattern is ϕ-polarized for even currents and θ-polarized for odd currents. The scale is linear, with each interval between tic marks representing an increment of two in gain. The gain patterns in other planes, such as the $x-y$ plane, are not simply related to those in the $x-z$ and $y-z$ planes. For complete information, some sort of three-dimensional presentation of the gain patterns would be desirable.

Modal solutions for wire antennas and wire scatterers have also been made and compared to the matrix inversion solution [3.11]. As modes were added, the modal solution converged to the matrix inversion solution in about the same way as for bodies of revolution. In the antenna problem, given a voltage source at some point along the wire, the modes are excited in proportion to their current amplitudes at the point of excitation. By using several voltage sources at several points, the excita-

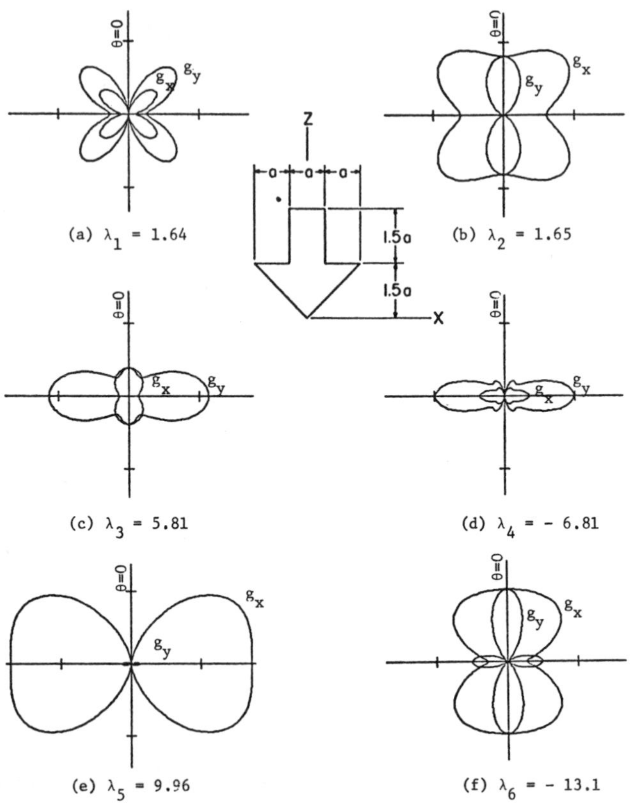

Fig. 3.6a–f. Characteristic gain patterns for the wire arrow, with "a" 0.25 wavelength and wire diameter 0.004 wavelength. Patterns in the $x=0$ plane are labeled g_x, those in the $y=0$ plane g_y. The six lowest-order modes are shown

tion of a number of modes could be completely controlled. In the scattering problem, the excitation of modes could be controlled by placing lumped loads along the wire. However, if the loads were restricted to be passive, only partial control of the mode excitation would be possible.

3.5. Control of Modes by Reactive Loading

It is well known that the radiation and scattering properties of a conducting body can be grossly changed by impedance loading [3.13, 14]. If the body is electrically small or intermediate in size, then only a few characteristic modes contribute to the radiation and scattering patterns. Hence, by controlling these modes one can control the electromagnetic behavior of a body. In this section methods for obtaining desired characteristic modes and for using them in synthesis problems are given.

3.5.1. Modes of a Loaded Body

A loaded body is defined as one for which the tangential electric field $\boldsymbol{E}_{\text{tan}}$ on its surface S is related to the current $\boldsymbol{J}$ on S by an impedance function Z_{L}, i.e.

$$\boldsymbol{E}_{\text{tan}} = Z_{\text{L}}(\boldsymbol{J}) . \tag{3.72}$$

The total electric field is the sum of the impressed field $\boldsymbol{E}^{\text{i}}$ plus the scattered field $\boldsymbol{E}^{\text{s}}$ produced by the current $\boldsymbol{J}$, i.e.,

$$\boldsymbol{E} = \boldsymbol{E}^{\text{i}} + \boldsymbol{E}^{\text{s}} . \tag{3.73}$$

As discussed in Section 3.1, the tangential component of $\boldsymbol{E}^{\text{s}}$ on S is related to the impedance operator $Z(\boldsymbol{J})$ for S by

$$\boldsymbol{E}^{\text{s}}_{\text{tan}} = -Z(\boldsymbol{J}) . \tag{3.74}$$

Hence, specializing (3.73) to tangential components on S, and substituting from (3.72) and (3.74), one has

$$(Z + Z_{\text{L}})(\boldsymbol{J}) = \boldsymbol{E}^{\text{i}}_{\text{tan}} . \tag{3.75}$$

This is the operator equation for the current $\boldsymbol{J}$ on a loaded surface S excited by an impressed field $\boldsymbol{E}^{\text{i}}$. It is of the same form as the equation for conducting surfaces (3.1) except that the operator $(Z + Z_{\text{L}})$ in (3.75) replaces the operator $L_{\text{tan}} = Z$ in (3.1).

Modal solutions to (3.75) are obtained by a procedure analogous to that for conducting surfaces (Section 3.1). Consider the eigenvalue equation

$$(Z + Z_{\text{L}})(\boldsymbol{J}_n) = v_n M(\boldsymbol{J}_n) , \tag{3.76}$$

where v_n are eigenvalues, $\boldsymbol{J}_n$ are eigenfunctions, and M is a weight operator to be chosen. In general,

$$Z = R + jX \quad \text{and} \quad Z_{\text{L}} = R_{\text{L}} + jX_{\text{L}} \tag{3.77}$$

where R and X are the Hermitian parts of Z, and R_{L} and X_{L} are the Hermitian parts of Z_{L}. If the loads are lossy, that is, if $R_{\text{L}} \neq 0$, one has two choices: a) Set $M = R + R_L$, in which case the modal currents are real, but orthogonality of the modal radiation fields is lost. b) Set $M = R$, in which case orthogonality of the modal radiation patterns is retained, but the modal currents may be complex.

In this section only reactive loads will be considered, that is, $R_L = 0$. This load is loss free, and the preceeding two choices for M become the same. Hence, let $M = R$ and $v_n = 1 + j\lambda_n$ in (3.76), cancel the common terms $R(\boldsymbol{J}_n)$, and obtain

$$(X + X_L)\,\boldsymbol{J}_n = \lambda_n\, R(\boldsymbol{J}_n)\,. \tag{3.78}$$

This is of the same form as for conducting bodies, (3.11), except that the operator $(X + X_L)$ in (3.78) replaces the operator X in (3.11). With this slight change, all of the properties of modal currents and modal fields remain the same as for conducting bodies. For example, the eigenvalues λ_n are all real, and the eigencurrents $\boldsymbol{J}_n$ can be chosen real. Orthogonality relationship (3.12) remains unchanged, and the X of (3.13) should be replaced by $X + X_L$. The orthogonality of radiation fields, (3.18), also remains unchanged. The use of characteristic modes for radiation and scattering problems for reactively loaded bodies remains the same as for conducting bodies (Section 3.2).

3.5.2. Resonating a Desired Real Current

The modal solution for the field radiated or scattered by a reactively loaded body continues to be given by (3.27), where the modal excitation coefficients are given by (3.25). The modes which contribute most to $\boldsymbol{E}$ are those having the smallest $|\lambda_n|$, under the assumption that they are excited. A mode having $|\lambda_n| = 0$ is said to be externally resonant. If no other $|\lambda_n|$ is small, then the radiated or scattered field consists almost entirely of the single resonant mode. This condition is usually met for bodies of electrically small or intermediate size.

For a reactively loaded body, any real current $\boldsymbol{J}$ can be made an eigencurrent corresponding to the eigenvalue $\lambda = 0$ by choosing the appropriate loading reactance. To illustrate, let

$$X_L(\boldsymbol{J}) = -\,X(\boldsymbol{J}) \tag{3.79}$$

whence the left-hand side of (3.78) is zero. If $\boldsymbol{J}$ is not associated with an internal resonance, then $R(\boldsymbol{J}) \neq 0$ and the eigenvalue λ must be zero. If X_L is a simple function of proportionality, then (3.79) is an equation for determining the function X. This general procedure will be called modal resonance.

The preceding procedure becomes clearer when expressed in matrix notation. To be explicit, consider a wire object and apply the method of moments with triangle functions for expansion and testing. Let s be the length variable along the wire, $I(s)$ the current on the wire, and $T(s-s_i)$ a triangle function extending from point s_{i-1} to point s_{i+1}, with unit peak at s_i. The current is expressed as

$$I(s) = \sum_i I_i \, T(s-s_i)\,, \tag{3.80}$$

where $I_i = I(s_i)$ are unknown coefficients. Applying the method of moments with testing functions equal to the expansion functions, one obtains the matrix equation

$$[Z + Z_{\mathrm{L}}]\,\bar{I} = \bar{V}^{\mathrm{i}}, \tag{3.81}$$

where $\bar{I}$ is the column vector of the I_i, and $\bar{V}^{\mathrm{i}}$ is the excitation vector with elements

$$V_j^{\mathrm{i}} = \int_{\text{wire}} T(s-s_i)\, E_s^{\mathrm{i}}(s)\, ds\,. \tag{3.82}$$

The matrix $[Z]$ is the generalized impedance matrix for the wire [3.9], and $[Z_{\mathrm{L}}]$ is the load impedance matrix with elements

$$(Z_{\mathrm{L}})_{ij} = \int_{\text{wire}} ds' \int_{\text{wire}} ds\, T(s'-s_i)\, Z_{\mathrm{L}}(s)\, T(s-s_j)\,. \tag{3.83}$$

Note that $[Z_{\mathrm{L}}]$ is at most a tridiagonal matrix, since Z_{L} is an ordinary function of position. To simplify computation, $Z_{\mathrm{L}}(s)$ is approximated by lumped inpedances at the points s_i, in which case $[Z_{\mathrm{L}}]$ becomes a diagonal matrix. Experience shows that this makes no noticeable difference compared to continuous loading so long as the s_i are close together.

For the present problem $[Z_{\mathrm{L}}]$ is a pure reactance $[jX_{\mathrm{L}}]$. Hence, the matrix representation of the eigenvalue equation (3.78) is

$$[X + X_{\mathrm{L}}]\,\bar{I}_n = \lambda_n [R]\,\bar{I}_n\,, \tag{3.84}$$

where $[X_{\mathrm{L}}]$ is a diagonal matrix with diagonal elements $X_{\mathrm{L}i}$. To resonate a real current $\bar{I}$, set the right-hand side of (3.84) equal to zero, and, analogous to (3.79), obtain

$$X_{\mathrm{L}i} I_i = -([X]\,\bar{I})_i \tag{3.85}$$

for $i = 1, 2 \ldots$. The notation $([X]\,\bar{I})_i$ means the ith element of the column vector $[X]\,\bar{I}$. Since $[X]$ and $\bar{I}$ (the desired current) are known, the reactive loads can be computed from (3.85).

3.5.3. A Scattering Example

For sample computations, take the thin wire triangle of Fig. 3.7, The total length of wire is $2L$ = one wavelength, and the wire diameter is $d = 0.01\,L$. The triangle lies in the $y = 0$ plane, with tip (30° angle) at the origin. The coordinates of a field point are the usual r, θ, ϕ spherical coordinates. All computations were made using 30 triangle functions for expansion and testing. Hence, the generalized impedance matrix for the wire was 30 by 30 in size. The four lowest-order modes of the unloaded scatterer have eigenvalues $\lambda_1 = -0.515$, $\lambda_2 = -5.46$, $\lambda_3 = 31.8$, $\lambda_4 = 135$, listed in order of increasing $|\lambda|$. The first modal current is similar to a sine wave, starting and ending at the tip. The second mode is similar to a cosine wave, the third is almost constant, and the fourth similar to two cycles of a sine wave. Plots of the modal currents, their gain patterns, and convergence of the modal solution are given in [3.15].

Next the same wire scatterer was loaded to resonate the $\lambda = 31.8$ mode current of the unloaded wire triangle. The reactances $X_{\mathrm{L}i}$ required at the thirty points s_i to resonate the current were computed according to (3.85). These loads were then added to the generalized impedance matrix, as shown in (3.81) and the modal currents and fields computed with available programs [3.11]. The modal current distributions on the wire did not change greatly from those of the unloaded wire, but their

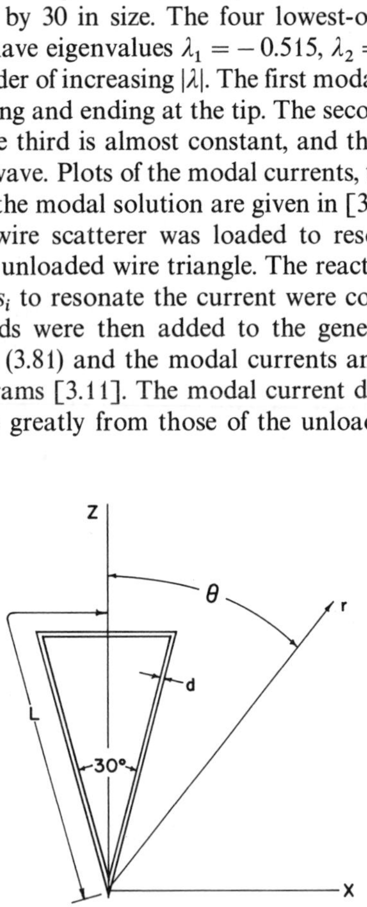

Fig. 3.7. Wire triangle used for computations ($2L$ = one wavelength, $d = 0.01\,L$)

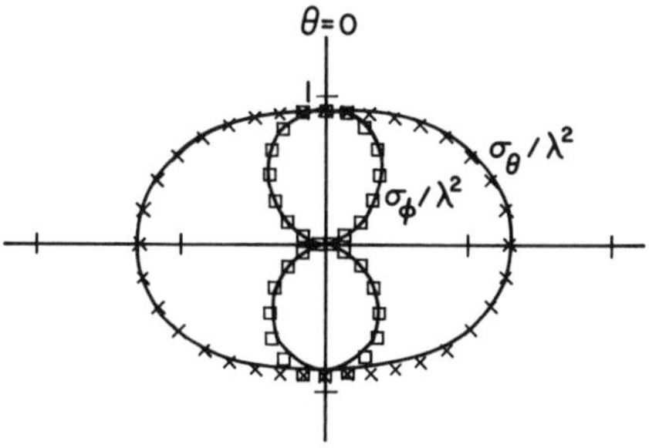

Fig. 3.8. Bistatic radar cross section/wavelength squared (σ/λ^2) for the reactively loaded wire triangle of Fig. 3.7, excited by a plane wave axially incident on the 30° angle. Solid curve shows the matrix inversion solution, crosses show the one-term modal solution in the E plane ($y=0$), and squares show the one-term modal solution in the H plane ($x=0$)

eigenvalues did. The new dominant mode had an eigenvalue of $\lambda=0$, of course. The other eigenvalues became $\lambda_2=-3.48$, $\lambda_3=-22.0$, and $\lambda_4=-175$. The modes of the loaded wire triangle were then used in a modal solution to a plane-wave scattering problem, and compared to the matrix inversion solution. Figure 3.8 shows the bistatic radar cross-section for a plane wave axially incident on the tip of the reactively loaded triangle. The one term modal solution ($\lambda=0$ mode only) in the E and H planes is shown by the crosses and squares, respectively. The corresponding matrix inversion solution is shown solid. Note that the two agree very closely, indicating that the scattering pattern consists almost entirely of the one mode resonated.

3.5.4. Synthesis of Loaded Scatterers

The modal resonance procedure allows one to resonate any desired real current, which can be chosen to synthesize loaded scatterers with desirable characteristics. A number of examples are given in [3.16]. One case, that of using reactive loads to broadband a scatterer, is summarized here.

It is often desirable to have a scatterer whose radar cross section changes slowly with frequency. As a measure of the frequency sensitivity of a current $\bar{I}$, one can define a quality factor as

$$Q=\omega\tilde{I}^*[X]\,\bar{I}/\tilde{I}^*[R]\,\bar{I}\,, \tag{3.86}$$

where $[X']=[dX/d\omega]$. If $\bar{I}$ is an eigencurrent, an interpretation can be made in terms of the frequency variation of the eigenvalue, as follows. Consider the Rayleigh quotient for λ, and assume that the dominant frequency variation is due to that of $[X]$. Taking the frequency derivative of the Rayleigh quotient, one has

$$Q \approx \omega\, \partial\lambda / \partial\omega \tag{3.87}$$

This relationship becomes exact at resonance, i.e., if $\lambda = 0$. A scatterer will be called broadband if only low Q modes contribute significantly to the scattered field.

The lowest Q currents for a given body can be found from (3.86) in the usual way by setting its first variation equal to zero. The resulting eigenvalue equation is

$$[\omega X']\, \boldsymbol{I} = Q[R]\, \bar{\boldsymbol{I}}\,. \tag{3.88}$$

where Q is the eigenvalue. The smallest eigenvalue is the minimum Q for all possible currents $\bar{\boldsymbol{I}}$. Equation (3.88) has been solved for the wire triangle of Fig. 3.7, and the smallest Q found to be $Q = 7.38$. Its current is an odd function about the midpoint of the length variable, and it radiates maximum field in the $z = 0$ plane. It will be called a broadside mode. The next lowest Q was found to be $Q = 27.6$. Its current is an even function about the midpoint of the length variable, and it radiates maximum field in the z direction. This current is called the lowest Q endfire current.

The lowest Q endfire current was next resonated according to the concepts of Subsection 3.5.2. The required load reactances were found from (3.85) and the modes of the loaded triangle calculated by the programs of [3.11]. The dominant mode current became the lowest Q endfire current, as expected. The other mode currents were all different from those of the unloaded triangle. Next the modal solution for the loaded wire was calculated and compared to the matrix inversion solution. Again a one-term modal solution was very close to the matrix inversion solution, showing that only one mode was contributing significantly to the scattering pattern. Also, the scattered field pattern was very similar to the field pattern of a magnetic dipole, which agrees with one's intuition that low Q currents are associated with dipole fields.

To illustrate the variation of radar cross-section σ over a frequency band, graphs of backscattering cross-section versus frequency are shown in Fig. 3.9. The object is a wire angle-circle (cross section of a cone-sphere) under several loading conditions. As illustrated in the insert of Fig. 3.9, the object consists of a 12° wire angle closed by a wire circle of 20 cm outside diameter. The wire diameter is one-tenth the circle diameter. This object is naturally resonant in the vicinity of 175 MHz. The wire is excited by a plane wave axially incident on the 12° angle.

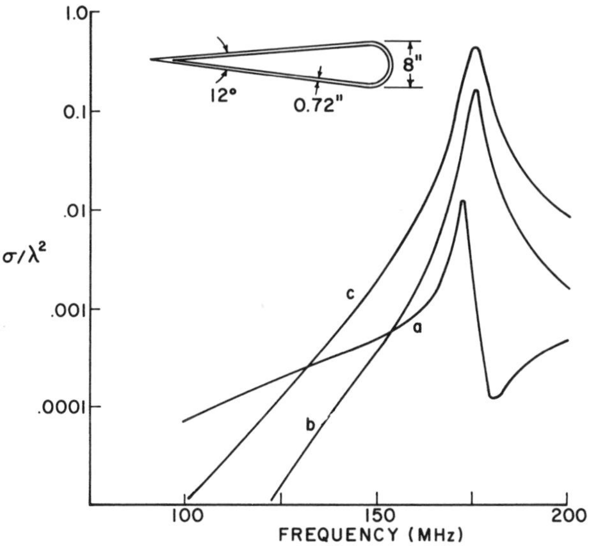

Fig. 3.9. Backscattering radar cross section/wavelength squared for the wire angle-circle (cross section of a cone-sphere). Curve *a*: unloaded wire. Curve *b*: wire loaded to resonate the most inductive eigencurrent of the unloaded wire. Curve *c*: wire loaded to resonate the lowest Q endfire current

Curve *a* shows radar cross-section per square wavelength (σ/λ^2) for the unloaded wire. Curve *b* shows σ/λ^2 for the wire loaded to resonate the eigencurrent with largest positive λ_n (the example of Subsection 3.5.3). Curve *c* exhibits σ/λ^2 for the wire loaded to resonate the lowest Q endfire current. Note that the lowest Q case is considerably more broadband than the other two.

3.6. Characteristic Modes for Dielectric Bodies

Characteristic modes can be used for dielectric bodies, for magnetic bodies, and for both those dielectric and magnetic. These bodies may be lossy or loss-free. The general case of bodies both dielectric and magnetic is considerably more complicated than that for bodies dielectric but non-magnetic, and hence only this latter case is treated here. The general case was developed in [3.17].

Consider a loss-free non-magnetic body, that is, one with permittivity ε real and possibly a function of position, and with permeability μ equal to the free-space value μ_0. The electric field is represented as the sum of an incident field $\boldsymbol{E}^{\mathrm{i}}$, due to impressed sources, plus a scattered field

$\boldsymbol{E}^s$, due to induced polarization current $\boldsymbol{J}$. The polarization current is related to the total field by the constitutive relationship

$$\boldsymbol{J} = j\omega(\varepsilon - \varepsilon_0)(\boldsymbol{E}^i + \boldsymbol{E}^s). \tag{3.89}$$

The scattered field is related to the polarization current by a linear operator Z_V, which depends only on the shape of V, according to

$$\boldsymbol{E}^s = -Z_V(\boldsymbol{J}). \tag{3.90}$$

In terms of the magnetic vector potential integral $\boldsymbol{A}$ and the electric scalar potential integral ϕ,

$$Z_V(\boldsymbol{J}) = j\omega\boldsymbol{A}(\boldsymbol{J}) + \nabla\phi(\boldsymbol{J}). \tag{3.91}$$

Substituting (3.90) into (3.89) and rearranging, one has

$$Z_V(\boldsymbol{J}) + (j\omega\Delta\varepsilon)^{-1}\boldsymbol{J} = \boldsymbol{E}^i, \tag{3.92}$$

where $\Delta\varepsilon = \varepsilon - \varepsilon_0$. This is an operator equation for the polarization current in dielectric bodies in an electromagnetic field.

To emphasize the analogy with the corresponding formulation for conducting bodies, introduce the impedance operator

$$Z = Z_V + (j\omega\Delta\varepsilon)^{-1} \tag{3.93}$$

and rewrite (3.92) as

$$Z(\boldsymbol{J}) = \boldsymbol{E}^i. \tag{3.94}$$

Also, define the symmetric product of two vector functions $\boldsymbol{B}$ and $\boldsymbol{C}$ in V as

$$\langle \boldsymbol{B}, \boldsymbol{C} \rangle = \iiint_V \boldsymbol{B} \cdot \boldsymbol{C} \, d\tau. \tag{3.95}$$

The product $\langle \boldsymbol{B}^*, \boldsymbol{C} \rangle$, where * denotes complex conjugate, defines an inner product for the Hilbert space of square integrable vector functions in V.

It follows from the reciprocity theorem that Z is a symmetric operator, that is, $\langle J_1, Z_V J_2 \rangle = \langle J_2, Z_V J_1 \rangle$. If $\Delta\varepsilon$ is a scalar or a symmetric tensor, $(j\omega\Delta\varepsilon)^{-1}$ is also a symmetric operator. Hence Z, given by (3.93), is symmetric, but not Hermitian. It can be expressed in terms of Hermitian parts as $Z = R + jX$, where

$$R = (Z + Z^*)/2, \tag{3.96}$$

$$X = (Z - Z^*)/2_j. \tag{3.97}$$

For loss-free dielectrics, R is also the real Hermitian part of Z_V, since $(j\omega\Delta\varepsilon)^{-1}$ is imaginary. The complex power balance for a current $\boldsymbol{J}$ in V is

$$\begin{aligned} P &= \langle \boldsymbol{J}^*, Z_V \boldsymbol{J} \rangle \\ &= \oint\!\!\oint_{S'} \boldsymbol{E} \times \boldsymbol{H}^* \cdot d\boldsymbol{s} + j\omega \iiint_{V'} (\mu_0 \boldsymbol{H} \cdot \boldsymbol{H}^* - \varepsilon_0 \boldsymbol{E} \cdot \boldsymbol{E}^*)\, d\tau, \end{aligned} \tag{3.98}$$

where S' is any surface surrounding V, and V' is the region enclosed by S'. The real part of $\langle \boldsymbol{J}^*, Z\boldsymbol{J} \rangle$ is

$$\langle \boldsymbol{J}^*, R\boldsymbol{J} \rangle = \mathrm{Re}\{\langle \boldsymbol{J}^*, Z_V \boldsymbol{J} \rangle\} = \mathrm{Re}\{P\} \tag{3.99}$$

which is the time-average power. The imaginary part of $\langle \boldsymbol{J}^*, Z\boldsymbol{J} \rangle$ is

$$\langle \boldsymbol{J}^*, X\boldsymbol{J} \rangle = \mathrm{Im}\{\langle \boldsymbol{J}^*, Z_V \boldsymbol{J} \rangle\} - \omega^{-1} \iiint_V \boldsymbol{J}^* \cdot (\Delta\varepsilon)^{-1} \boldsymbol{J}\, d\tau \tag{3.100}$$

which is not the imaginary part of P.

The eigenvalue equation defining the characteristic modes is

$$X(\boldsymbol{J}_n) = \lambda_n R(\boldsymbol{J}_n) \tag{3.101}$$

where X and R are real symmetric operators. Therefore all eigenvalues λ_n are real and all characteristic currents $\boldsymbol{J}_n$ can be chosen real. The $\boldsymbol{J}_n$ can be normalized to radiate unit power, and they satisfy orthogonality relationships

$$\langle \boldsymbol{J}_m^*, R\boldsymbol{J}_n \rangle = \delta_{mn}, \tag{3.102}$$

$$\langle \boldsymbol{J}_m^*, X\boldsymbol{J}_n \rangle = \lambda_n \delta_{mn}, \tag{3.103}$$

$$\langle \boldsymbol{J}_m^*, Z\boldsymbol{J}_n \rangle = (1 + j\lambda_n)\, \delta_{mn}. \tag{3.104}$$

The electric field $\boldsymbol{E}_n$ and the magnetic field $\boldsymbol{H}_n$ produced by a characteristic current $\boldsymbol{J}_n$ are called characteristic fields. If V is of finite extent, the characteristic fields in the radiation zone are outward traveling waves.

The characteristic far-fields satisfy the orthogonality relationship

$$(\varepsilon/\mu)^{1/2} \iint_{S_\infty} \boldsymbol{E}_m^* \cdot \boldsymbol{E}_n \, ds = \delta_{mn} , \tag{3.105}$$

where S_∞ is the sphere at infinity. Modal solutions exist for $\boldsymbol{J}$ of the form

$$\boldsymbol{J} = \sum_n V_n^{\mathrm{i}} \boldsymbol{J}_n (1 + j\lambda_n)^{-1} , \tag{3.106}$$

where the V_n^{i} are mode excitation coefficients

$$V_n^{\mathrm{i}} = \iiint_V \boldsymbol{J}_n \cdot \boldsymbol{E}^{\mathrm{i}} \, d\tau . \tag{3.107}$$

Similar modal solutions exist for the fields $\boldsymbol{E}^{\mathrm{s}}$ and $\boldsymbol{H}^{\mathrm{s}}$. The characteristic fields diagonalize the scattering and the perturbation matrices for dielectric bodies, in the same way as for conducting bodies (Section 3.3).

3.7. Characteristic Modes for N-Port Loaded Bodies

An N-port loaded body is one having N ports (or terminal pairs) to which N impedance loads (or an N-port network) is connected [3.13, 18]. The electromagnetic characteristics of the system depend on the load network as well as on the body itself. In this section the problem will be treated as a scattering problem, although the same theory applies to antenna systems. The analysis can be applied either in terms of impedance parameters or admittance parameters. For this section only the impedance formulation will be discussed. The admittance formulation is given in [3.19].

3.7.1. Formulation of the Problem

A loaded scatterer is basically two N-port networks connected together. The load network is passive, and its terminal characteristics can be represented by its N-port impedance matrix $[Z_{\mathrm{L}}]$. The scatterer, illuminated by an impressed electric field $\boldsymbol{E}^{\mathrm{i}}$, is an active network and its terminal characteristics can be represented by a Thévenin equivalent circuit [3.20]. This equivalent consists of an N-port impedance matrix $[Z_{\mathrm{S}}]$ obtained by removing the excitation (impressed field), plus series voltage sources at each port equal to the open-circuit port voltages V_n^{oc} which exist when all ports are open circuited. Figure 3.10 shows this Thévenin equivalent connected to the load network. The terminal

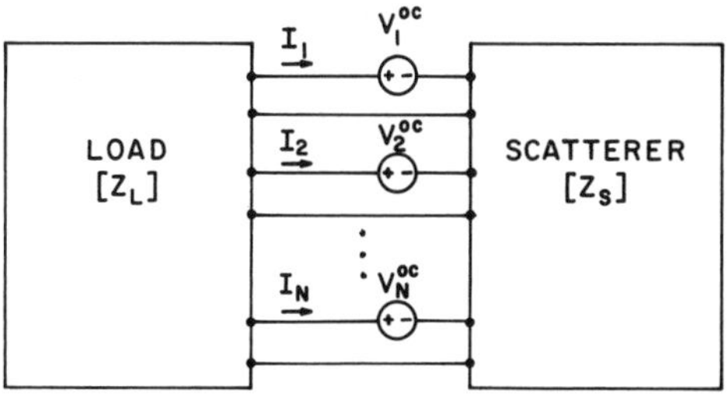

Fig. 3.10. Thévenin equivalent of the illuminated scatterer connected to a load network

equation is

$$\bar{V}^{oc} = -[Z_S + Z_L]\bar{I}, \tag{3.108}$$

where $\bar{V}^{oc}$ and $\bar{I}$ are the column matrices (vectors) of the voltage sources and port currents, respectively, with reference conditions as shown in Fig. 3.10. The reference condition for current is chosen to be into the positive voltage terminals for the scatterer and out of the positive voltage terminals for the load.

The scattered electric field $\boldsymbol{E}^s$ can be written as the superposition

$$\boldsymbol{E}^s = \boldsymbol{E}_0^{oc} + \sum_{n=1}^{N} \tilde{I}_n \boldsymbol{E}_n^{oc}. \tag{3.109}$$

Here $\boldsymbol{E}_0^{oc}$ is the field scattered when all ports are open circuited, and $\boldsymbol{E}_n^{oc}$ is the field radiated when a unit current exists at port n and all other ports are open circuited. In matrix form (3.109) can be written as

$$\boldsymbol{E}^s = \boldsymbol{E}_0^{oc} + \tilde{\boldsymbol{E}}^{oc}\bar{I} \tag{3.110}$$

where $\tilde{\boldsymbol{E}}^{oc}$ is the row matrix of the $\boldsymbol{E}_n^{oc}$, $n = 1, 2, \ldots, N$. Equation (3.108) can be solved for the port currents as

$$\bar{I} = -[Z_S + Z_L]^{-1}\,\bar{V}^{oc}. \tag{3.111}$$

Substituting this into (3.110), one obtains

$$\boldsymbol{E}^s = \boldsymbol{E}_0^{oc} - \tilde{\boldsymbol{E}}^{oc}[Z_S + Z_L]^{-1}\,\bar{V}^{oc}. \tag{3.112}$$

The current $\boldsymbol{J}$ on the scatterer must also be a superposition of the form (3.112), or

$$\boldsymbol{J} = \boldsymbol{J}_0^{oc} - \tilde{\boldsymbol{J}}^{oc}[Z_S + Z_L]^{-1}\,\bar{V}^{oc}. \tag{3.113}$$

Here $\boldsymbol{J}_0^{\mathrm{oc}}$ is the current induced on the scatterer by $\boldsymbol{E}^{\mathrm{i}}$ when all ports are open circuited, and $\tilde{\boldsymbol{J}}^{\mathrm{oc}}$ is the row matrix with elements $\boldsymbol{J}_n^{\mathrm{oc}}$, the current on the scatterer when a unit current exists at port n and all other ports are open circuited.

3.7.2. *N*-Port Characteristic Modes

The characteristic modes for a loaded N-port system are defined in a manner analogous to those for continuously loaded bodies (Section 3.5). Both $[Z_{\mathrm{S}}]$ and $[Z_{\mathrm{L}}]$ are assumed symmetric. Hence their sum is symmetric and can be expressed in terms of real and imaginary parts as

$$[Z] = [Z_{\mathrm{S}} + Z_{\mathrm{L}}] = [R] + j[X], \tag{3.114}$$

where

$$[R] = [Z + Z^*]/2, \tag{3.115}$$

$$[X] = [Z - Z^*]/2j. \tag{3.116}$$

The characteristic modes of the N-port system are defined by the weighted eigenvalue equation

$$[Z]\,\bar{\boldsymbol{I}}_n = (1 + j\lambda_n)\,[R]\,\bar{\boldsymbol{I}}_n, \tag{3.117}$$

where $\bar{\boldsymbol{I}}_n$ are the eigenvectors and $(1 + j\lambda_n)$ are the eigenvalues. Substituting from (3.114) into (3.117), and cancelling the common $[R]\,\bar{\boldsymbol{I}}_n$ terms, one has

$$[X]\,\bar{\boldsymbol{I}}_n = \lambda_n [R]\,\bar{\boldsymbol{I}}_n. \tag{3.118}$$

This is a real symmetric eigenvalue equation. Hence, all eigenvalues λ_n are real and all eigenvectors $\bar{\boldsymbol{I}}_n$ may be chosen real. More generally, the $\bar{\boldsymbol{I}}_n$ are equiphasal, that is, a real vector times a complex constant. In (3.117) note that $[R]\,\bar{\boldsymbol{I}}_n$ is real, and $(1 + j\lambda_n)$ is just a complex constant. Hence, $[Z]\,\bar{\boldsymbol{I}}_n$, which can be viewed as the voltage sources of Fig. 3.10 which produce the mode currents, are also equiphasal.

The matrix $[R]$ is normally positive definite, since $\tilde{\boldsymbol{I}}^*[R]\,\bar{\boldsymbol{I}}$ is the time-average power radiated and/or dissipated by the system. For convenience, we normalize the mode currents so that they deliver unit power, that is,

$$\tilde{\boldsymbol{I}}_n^*\,[R]\,\bar{\boldsymbol{I}}_n = 1. \tag{3.119}$$

The usual equations of orthonormality of eigenvectors can now be written as

$$\tilde{\boldsymbol{I}}_m^* [R] \boldsymbol{\bar{I}}_n = \delta_{mn} , \tag{3.120}$$

$$\tilde{\boldsymbol{I}}_m^* [X] \boldsymbol{\bar{I}}_n = \delta_{mn} \lambda_n , \tag{3.121}$$

$$\tilde{\boldsymbol{I}}_m^* [Z] \boldsymbol{\bar{I}}_n = \delta_{mn} (1 + j\lambda_n) . \tag{3.122}$$

Because the $\boldsymbol{\bar{I}}_n$ are real or equiphasal, the orthogonality relationships remain valid even without conjugation of the first vector. For loss-free systems, all power input is radiated, and (3.120) leads to an orthogonality relationship for radiation fields, similar to that for conducting bodies (Subsection 3.1.2).

3.7.3. Modal Solutions

Modal solutions for the port currents are obtained in the usual way by using the mode currents as a basis for the port currents. To be explicit, let the port current be represented as

$$\boldsymbol{\bar{I}} = \sum_{n=1}^{N} \alpha_n \boldsymbol{\bar{I}}_n , \tag{3.123}$$

where the α_n are coefficients to be determined. Substitute (3.123) into (3.108) and obtain

$$\boldsymbol{\bar{V}}^{\mathrm{oc}} = - \sum_{n=1}^{N} \alpha_n [Z_{\mathrm{S}} + Z_{\mathrm{L}}] \boldsymbol{\bar{I}}_n . \tag{3.124}$$

Now take the scalar product of (3.124) with each $\tilde{\boldsymbol{I}}_m^*$ in turn to obtain the set of equations

$$\tilde{\boldsymbol{I}}_m^* \boldsymbol{\bar{V}}^{\mathrm{oc}} = - \sum_{n=1}^{N} \alpha_n \tilde{\boldsymbol{I}}_m^* [Z_{\mathrm{S}} + Z_{\mathrm{L}}] \boldsymbol{\bar{I}}_n \tag{3.125}$$

$m = 1, 2, \ldots, N$. Because of the orthogonality relationship (3.122), only the $n = m$ term remains, and (3.125) reduces to

$$\tilde{\boldsymbol{I}}_n^* \boldsymbol{\bar{V}}^{\mathrm{oc}} = - \alpha_n (1 + j\lambda_n) . \tag{3.126}$$

Substituting these values of α_n into (3.123), one has the modal solution for the port currents

$$\bar{I} = -\sum_{n=1}^{N} \tilde{I}_n \bar{V}^{oc} \bar{I}_n (1 + j\lambda_n)^{-1} . \tag{3.127}$$

If the mode currents are not normalized, the factors $1 + j\lambda_n$ should be replaced by $(1 + j\lambda_n) \tilde{I}_n^* [R] \bar{I}_n$.

That part of the scattered field controlled by the load can also be expressed in spectral form. The total scattered field is obtained by substituting (3.127) into (3.110), giving

$$\boldsymbol{E}^s = \boldsymbol{E}_0^{oc} - \sum_{n=1}^{N} \tilde{I}_n^* \bar{V}^{oc} \boldsymbol{E}(\bar{I}_n) (1 + j\lambda_n)^{-1} . \tag{3.128}$$

Here $\boldsymbol{E}(\bar{I}_n)$ is the field radiated when $\bar{I}_n$ exists at the scatterer ports. A modal solution for the total current on the scatterer is obtained in the same way. The result is

$$\boldsymbol{J} = \boldsymbol{J}_0^{oc} - \sum_{n=1}^{N} \tilde{I}_n^* \bar{V}^{oc} \boldsymbol{J}(\bar{I}_n) (1 + j\lambda_n)^{-1} . \tag{3.129}$$

Here $\boldsymbol{J}(\bar{I}_n)$ is the current on the scatterer when $\bar{I}_n$ exists at its ports. Again, if unnormalized mode currents are used, the factors $1 + j\lambda_n$ in (3.128) and (3.129) should be replaced by $(1 + j\lambda_n) \tilde{I}_n^* [R] \bar{I}_n$.

3.7.4. Modal Resonance

The mode currents $\bar{I}_n$ can be viewed as being excited by the voltage sources $V^{oc} = -[Z] \bar{I}_n$ in the circuit of Fig. 3.10. The power delivered by the sources is

$$P = \tilde{I}_n^* [Z] \bar{I}_n = 1 + j\lambda_n , \tag{3.130}$$

where the last equality follows from (3.122). A mode current is said to be in resonance when its eigenvalue λ_n is zero. Hence, at resonance, the reactive power λ_n is zero and the driving voltage is in phase with the current.

Any equiphase current vector $\bar{I}$ can be resonated by choosing the proper load $[Z_L]$. This means that a desired $\bar{I}$ can be made an eigencurrent with eigenvalue $\lambda = 0$. We see from (3.118) that this requires

$$[X] \bar{I} = [X_S + X_L] \bar{I} = 0 \tag{3.131}$$

or

$$[X_L]\,\bar{I} = -[X_S]\,\bar{I}\,. \tag{3.132}$$

This condition can always be satisfied by a diagonal load matrix with diagonal elements X_{Li}. Explicitly, the solution is then

$$X_{Li} = ([X_S]\,\bar{I})_i/I_i\,, \tag{3.133}$$

where $([X_S]\,\bar{I})_i$ denotes the ith component of the column matrix $[X_S]\,\bar{I}$. Loads more complicated than lumped loads can also be used and perhaps the non-uniqueness of the solution can be used for optimization purposes.

3.7.5. Synthesis of Loaded N-Port Scatters

Specialization of the modal solution to both near-field problems and to plane-wave scattering is given in [3.19]. A numerical method of solution is also discussed, and a number of examples of modal analysis given for a loaded wire triangle with two cross wires. The solution can be used for synthesis problems in the same manner, as described for continuously loaded bodies in Subsection 3.5.4. A number of examples of pattern synthesis are given in [3.21]. The synthesis procedure can synthesize a pattern specified in both magnitude and phase, or one specified in magnitude only. The latter case is a problem of mixed antenna synthesis, as defined by BAKHRAKH and TROYTSKIY [3.22]. The method of solution is similar to that used by CHONI [3.23]. The procedure is an iterative one in which the phase of the desired field is first chosen arbitrarily and the source obtained by a conventional synthesis procedure. The phase of the desired field is then changed to the phase of the synthesized field, and the synthesis procedure rerun. The procedure eventually converges because no step can increase error. It converges to a stationary point, usually a local minimum for the error, which may or may not be a global minimum.

3.8. Discussion

Characteristic modes are of value, both theoretically and computationally, for radiation and scattering problems for the following reasons: a) They diagonalize the generalize impedance matrix of the body, obviating the necessity of matrix inversion in a moment solution. b) They diagonalize the scattering matrix, obviating the necessity of matrix inversion in a radiation or scattering pattern synthesis procedure. c) Modal solutions for small and intermediate size bodies converge rapidly and in a least-squares sense on the radiation sphere. d) The characteristic modes can be controlled by impedance loads according to the modal resonance concept. This chapter summarizes a number of uses of the characteristic modes, illustrating the above listed properties.

Computer programs are available for all of the numerical examples given and discussed in this chapter. Listings of the programs, with documentation and sample input-output data, are available in a series of reports published by Syracuse University under contract to the Air Force Cambridge Research Laboratories [3.24]. Some of these programs have been deposited with the National Auxiliary Publication Service [3.25, 26], and it is planned to deposit others in the future. The programs are general purpose and user oriented, so that they may be used for a wide variety of problems by interested persons.

References

3.1. R.J. GARBACZ: "A General Expansion for Radiated and Scattered Fields"; Ph.D. Dissertation, Ohio State University, Columbus, Ohio (1968).

3.2. R.J. GARBACZ, R.H. TURPIN: IEEE Trans. Antennas Propagation AP-**19**, 348 (1971).

3.3. R.F. HARRINGTON, J.R. MAUTZ: IEEE Trans. Antennas Propagation AP-**19**, 622 (1971).

3.4. R.F. HARRINGTON: *Field Computation by Moment Methods* (The Macmillan Company, New York, 1968).

3.5. R.F. HARRINGTON: *Time-Harmonic Electromagnetic Fields* (McGraw-Hill, New York, 1961).

3.6. C. G. MONTGOMERY, R. H. DICKE, E. M. PURCELL: *Principles of Microwave Circuits*, Rad. Lab. Series (McGraw-Hill, New York, 1948).

3.7. R.F. HARRINGTON, J.R. MAUTZ: IEEE Trans. Antennas Propagation AP-**19**, 629 (1971).

3.8. J.R. MAUTZ, R.F. HARRINGTON: Appl. Sci. Res. **20**, 405 (1969).

3.9. D.C. KUO, H.H. CHAO, J.R. MAUTZ, B.J. STRAIT, R.F. HARRINGTON: IEEE Trans. Antennas Propagation AP-**20**, 814 (1972).

3.10. J.R. MAUTZ, R.F. HARRINGTON: "Computer Programs for Characteristic Modes of Bodies of Revolution"; Sci. Report No. 10, Contract No. F19628-68-C-0180, AFCRL, DDC No. AD 729-926 (August 1971).

3.11. J.R. MAUTZ, R.F. HARRINGTON: "Computer Programs for Characteristic Modes of Wire Objects"; Sci. Report. No. 11, Contract No. F19628-68-C-0180, AFCRL, DDC No. AD 722-062 (March 1971).

3.12. J.H. WILKINSON: *The Algebraic Eigenvalue Problem* (Oxford Press, London, 1967), p. 34.

3.13. J.K. SCHINDLER, R.B. MACK, P. BLACKSMITH, JR.: Proc. IEEE **53**, 993 (1965).

3.14. R.F. HARRINGTON, J.R. MAUTZ: Appl. Sci. Res. **26**, 209 (1971).

3.15. J.R. MAUTZ, R.F. HARRINGTON: "Control of Radar Scattering by Reactive Loading"; Sci. Report No. 13, Contract No. F19628-68-C-0180, AFCRL, DDC No. AD 729-926 (August 1971).

3.16. R. F. Harrington, J. R. Mautz: IEEE Trans. Antennas Propagation AP-**20**, 446 (1972).
3.17. R. F. Harrington, J. R. Mautz, Yu Chang: IEEE Trans. Antennas Propagation AP-**20**, 194 (1972).
3.18. R. F. Harrington: Proc. IEE (London) **111**, 617 (1964).
3.19. J. R. Mautz, R. F. Harrington: IEEE Trans. Antennas Propagation AP-**21**, 188 (1973).
3.20. H. J. Carlin, A. B. Giordano: *Network Theory* (Prentice Hall, Englewood Cliffs, N.J., 1964).
3.21. R. F. Harrington, J. R. Mautz: IEEE Trans. Antennas Propagation AP-**22**, 184 (1974).
3.22. L. D. Bakhrakh, V. I. Troytskiy: Radio Eng. and Electronic Phys. (English translation) **12**, 404 (1967).
3.23. Y. Choni: Radio Eng. and Electronic Phys. (English translation) **16**, 770 (1971).
3.24. R. F. Harrington et al.: Several Scientific Reports prepared under Contract No. F19628-68-C-0180 (1968–1973).
3.25. D. C. Kuo, H. H. Chao, J. R. Mautz, B. J. Strait, R. F. Harrington: IEEE Trans. Antennas Propagation AP-**20**, 814 (1972).
3.26. J. R. Mautz, R. F. Harrington: IEEE Trans. Antennas Propagation AP-**22**, 630 (1974).

4. Some Computational Aspects of Thin-Wire Modeling

E. K. MILLER and F. J. DEADRICK

With 8 Figures

Some of the computational aspects which may affect the validity and applicability of a numerical solution for a thin-wire structure are considered in this paper. These include: 1) structure segmentation, 2) current expansions, 3) the thin-wire approximation, 4) matrix factorization roundoff error, 5) near-field numerical anomalies, 6) multiple junction treatment, 7) wire-grid modeling, and 8) computer time required. The discussion will be based upon results obtained from a subsectional-collocation and point-matching solution to the thin-wire integral equation, but the implications which arise are of a more general nature. Minimizing the possibility deterious impact of the above on performing practical calculations will also be discussed.

4.1. Introduction

Some of the computational aspects which may affect the validity and practical applicability of a numerical solution for a wire structure obtained from a moment method treatment are considered here. The results presented, unless otherwise indicated, are obtained from a subsectional collocation solution using point matching and three-term current expansion of the thin wire electric-field integral equation (see Line 7 of Table 4.2)

$$\begin{aligned} \boldsymbol{E}^{\mathrm{I}}(r) \cdot \hat{\boldsymbol{t}}(r) = (j\omega\mu_0/4\pi) \int_{C(\boldsymbol{r})} I(\boldsymbol{r}') \{\hat{\boldsymbol{t}}(\boldsymbol{r}) \cdot \hat{\boldsymbol{t}}(\boldsymbol{r}') \\ + 1/k^2[(\hat{\boldsymbol{t}}(\boldsymbol{r}) \cdot \nabla)]\,[(\hat{\boldsymbol{t}}(\boldsymbol{r}') \cdot \nabla)]\}\, g\, ds', \end{aligned} \tag{4.1}$$

where

$$g = \mathrm{e}^{-jkR}/\mathrm{R}$$

$$\mathrm{k} = \omega\sqrt{\mu_0 \varepsilon_0}$$

and

$$R = |\boldsymbol{r} - \boldsymbol{r}' + a(\boldsymbol{r}')|$$

with $\hat{\boldsymbol{t}}(\boldsymbol{r})$ the tangent vector to the wire at $\boldsymbol{r}$, $\boldsymbol{E}^{\mathrm{I}}$ the incident field, $a(\boldsymbol{r}')$ the wire radius at $\boldsymbol{r}'$ in the direction $\hat{\boldsymbol{t}}(\boldsymbol{r}') \times (\boldsymbol{r} - \boldsymbol{r}')$ and the (suppressed) time variation $e^{j\omega t}$. The reader is referred to Harrington [4.1] for the general approach and to Poggio and Miller [4.2] for specific details as well as to Miller and Deadrick [4.3] where the material presented here is discussed more fully. Where possible, the findings which result, from this particular solution procedure are generalized to permit identification of the broader implications pertaining to similar numerical methods. Recommendations for avoiding possible pitfalls and more fully realizing the potential of such methods are also discussed.

It is worthwhile mentioning that (4.1) can be put in a number of other equivalent forms, by suitable manipulation. Integration by parts can be used for example to transfer one of the derivatives to the current. This form of the integral equation, which we refer to as the vector-scalar potential type, has been discussed by Harrington [4.1] and employed by Chao and Strait [4.4]. Another type is obtained by moving the differential operator outside the integral and solving this new form as a differential equation, a procedure which leads to the vector potential, or Hallen type of integral equation [4.5]. The Hallen equation has the advantage that it has the lowest-order integrand singularity of the three, but it can not be as easily applied to the more general geometries that the other two can be readily used for.

A moment-method solution of (4.1) (or the other equivalent forms) encompasses many factors in the course of reducing the original integral equation to a linear system or matrix form. What is basically involved, whatever the specific bases and weight functions employed, is a numerical characterization of the current distribution in terms of N sampled current values and computation of the associated tangential fields at M points on the structure where in general $M \geqq N$. When point matching is employed, for example, $N \leqq M \leqq N+1$. (See Table 4.2 below.) On the other hand, the use of Galerkin's method, whereby the tangential field is integrated, is essentially equivalent to using many more field values than unknowns, i.e., $M \gg N$, with N equations finally being obtained by summing (integrating) the field values in a suitable fashion. The actual implementation of Galerkin's method may be achieved using a rather coarse integration scheme, so that $M \gtrsim 2N$, e.g. [4.4]. Under certain circumstances the field may succumb to analytic integration, [4.6] but this is not the usual case. Some of the various basis and weight function combinations which have been employed or which appear feasible are discussed in more detail below.

An additional factor which must be considered in solving an electromagnetic problem via a moment-method solution of an integral equation is the error which is associated with the numerical result. It is possible to identify two essentially independent errors: 1) the physical modeling error, ε_p, which results from approximating the actual structure with a (usually) simplified configuration for computational purposes; and 2) a numerical calculation error, ε_N, which represents the difference between the computed results and an exact solution for the configuration selected. It is usually necessary to conduct experiments to evaluate ε_p, since the actual structure is rarely amenable to rigorous numerical treatment. On the other hand, ε_N may be estimated by comparing numerical results as the number of current samples is successively increased to obtain the convergence rate of the numerical solution.

Sone of the various aspects to be considered here include: a) current expansions; b) structural segmentation and boundary condition matching; c) multiple junction treatment; d) thin-wire approximation; e) matrix factorization round-off error; f) near-field behavior; g) wire grid modeling; and h) computer time requirements. Each of these items is discussed in turn below. The order of presentation is essentially that in which these various factors are encountered in an actual calculation.

4.2. Current Expansions

The basis function used for the current expansion can strongly affect both the solution accuracy and efficiency. Some of the factors related to the basis function are: 1) the number of current samples or unknowns required to obtain the current distribution to a desired numerical accuracy; 2) the computational effort necessary to evaluate the associated field values (or impedance matrix coefficients); 3) the number of field values required to match the boundary conditions to the desired accuracy; and 4) the integral equation type (or field expression) used. As will be further commented on below, items 1) and 3) are not unrelated.

Current bases can be classified: 1) as entire domain, defined over the entire (continuous) domain $C(\boldsymbol{r})$ of the integral operator; or 2) sub-domain, defined over a sub-section of $C(\boldsymbol{r})$. An example of the former is a Fourier series current expansion employed for a linear dipole [4.7], and of the latter are the piece-wise sinusoidal and piece-wise linear expansions used, respectively, by RICHMOND [4.6] and by CHAO and STRAIT [4.4]. Generally speaking, a current basis which most closely resembles the numerically convergent current will minimize ε_N for a given number N of unknowns, or conversely achieve a specified accuracy with a minimum value for N. The complete domain and sub-domain bases are discussed in order below.

4.2.1. Complete Domain Expansions

We summarize here some of the kinds of complete domain and sub-domain current expansions which have been or might be employed for wire antennas analysis. Table 4.1 shows complete domain representations based on Fourier, MacLauren, Chebyshev, Hermite, and Legendre polynomial series [4.7]. Results obtained by RICHMOND from using these expansions for analyzing the current excited on a linear dipole by a plane wave at broadside incidence are also given in Table 4.1. RICHMOND commented that the expansion based on the Chebyshev and Legendre polynomials may be most promising for further investigation. For many reasons, some of which are given below, complete domain current

Table 4.1. Polynomial series current expansions. (After RICHMOND [4.7])

Fourier:

$$I(z) = I_1 \cos(\pi x/2) + I_2 \cos(3\pi x/2) + I_3 \cos(5\pi x/2) \quad + \quad +$$

MacLaurin:

$$I(z) = I_1 \qquad + I_2 x^2 \qquad + I_3 x^4 \qquad + \quad +$$

Chebyshev:

$$I(z) = I_1 T_0(x) + I_2 T_2(x) + I_3 T_4(x) \quad + \quad +$$

Hermite:

$$I(z) = I_1 H_0(x) + I_2 H_2(x) + I_3 H_4(x) \quad + \quad +$$

Legendre:

$$I(z) = I_1 P_0(x) + I_2 P_2(x) + I_3 P_4(x) \quad + \quad +$$

where

$$-1/2 \leqq x = 2z/L \leqq 1/2 .$$

Coefficients for the current $I(z)$ for an incident plane wave:

$$L = 0.5\,\lambda; \quad a = 0.005\,\lambda; \quad \theta_i = 90°$$

and,

T_i, H_i, and P_i represent, respectively the Chebyshev, Hermite, and Legendre functions of order i.

I_n	Fourier	MacLaurin	Chebyshev	Hermite	Legendre
1	3.476	3.374	1.7589	8.2929	2.2763
2	0.170	4.037	1.5581	14.3644	2.1005
3	0.085	3.128	0.0319	4.4135	0.0655
4	0.055	4.101	0.0112	0.3453	0.0421
5	0.040	1.871	0.0146	0.0073	0.0372

representations have not been extensively employed in wire antenna theory.

Sub-domain representations seem to be more popular because they are apparently easier to use and generally speaking provide comparable accuracy with shorter calculation times. One possible explanation for this situation is that a complete domain representation in a series form such as illustrated in Table 4.1 requires integration over the entire wire structure for each of the N terms in the series. The sub-domain representation, on the other hand, involves integration of the term or terms used for the current representation over only a small part (one segment) of the entire structure. Furthermore, by choosing the appropriate kinds of expansion functions it may be possible to perform such integrations analytically. Finally, the likelihood of encountering an ill-conditioned matrix is less when using a sub-domain basis. Some of the general features of sub-domain expansions are discussed below.

4.2.2. Sub-Domain Expansions

A number of sub-domain bases and weight function combinations are summarized in Table 4.2. Included here are examples which have already been implemented, as well as some which appear feasible, but which have not apparently been used. Included in the table are: 1) a word description of the method used, 2) the integral equation type which has been employed and references where appropriate, 3) the basis function in terms of the coefficients associated with it for segment n, 4) the current conditions employed (if any) to reduce the number of unknowns to $\sim N$; 5) the final form of the basis function written in terms of the sampled current values, 6) the weight function used, written as associated with segment m, 7) the final number of unknowns which result for an open-ended single wire and a loop (multiple junctions are not included here but are considered in Section 4.4 later); and 8) comments concerning the method's use. Note that the two-term linear and sinusoidal bases can also be viewed as overlapping functions. Also observe that for generality we allow the segment lengths, denoted by Δ, to be unequal. Finally, where the conditions differ between segments at open ends and interior segments, separate results are given for each. For simplicity we assume the segments to be numbered sequentially from 1 to N in the direction of both the increasing source coordinate and current reference.

Of the preceeding methods, 1), 2), 4), 5), and 7) have probably been most extensively employed. Note that all of the higher order (non-pulse) bases represent attempts to smooth or interpolate the current between its sampled values. Their relative success in doing this may be measured by how few samples can be used to achieve a desired accuracy and how

Table 4.2. Basis and weight function treatments

$I_{n,n+1}$ denotes current at junction of segments n and $n+1$ — $I'_{n,n+1}$ denotes the current derivative between s_n and s_{n+1}

Method	Integral equation type	Basis function	Current conditions: Interior segment	End segment (if different)
1 Pulse with point matching	Pocklington (RICHMOND [4.7])	I_n	None	—
2 Pulse with point matching	Vector-scalar potential (HARRINGTON [4.1])	I_n	None specifically on current. Current derivative (charge) $I'_{n,n+1} \equiv (I_{n+1} - I_n)/[(\Delta_n + \Delta_{n+1})/2]$	$I'_{0,1} = I_1/\Delta_1$ $I'_{N,N+1} = I_N/\Delta_N$
3 Piecewise sinusoidal with point matching	Pocklington (TAYLOR [4.26])	$[A_n^{(+)}\sin(ks_{n+})$ $+ A_n^{(-)}\sin(ks_{n-})]/\sin(k\Delta_n)$	$A_n^{(+)} = A_{n+1}^{(-)} = I_{n,n+1}$	$A_N^{(+)} = A_1^{(-)} = 0$
4 Piecewise linear, Galerkin's	Vector-scalar potential (CHAO and STRAIT [4.4])	$[A_n^{(+)}s_{n+} + A_n^{(-)}s_{n-}]/\Delta_n$	$A_n^{(+)} = A_{n+1}^{(-)} = I_{n,n+1}$	$A_N^{(+)} = A_1^{(-)} = 0$
5 Piecewise sinusoidal, Galerkins	Pocklington (RICHMOND [4.6])	$[A_n^{(+)}\sin(ks_{n+})$ $+ A_n^{(-)}\sin(ks_{n-})]/\sin(k\Delta_n)$	$A_n^{(+)} = A_{n+1}^{(-)} = I_{n,n+1}$	$A_N^{(+)} = A_1^{(-)} = 0$
6 Three-term algebraic with point matched weights and extrapolated currents	Pocklington (NEUREUTHER et al. [4.8])	$A_n + B_n s'_n + C_n s'^2_n$	$A_n - B_n\Delta_{n-1,n} + C_n\Delta^2_{n-1,n} = I_{n-1}$ $A_n = I_n$ $A_n + B_n\Delta_{n,n+1} + C_n\Delta^2_{n,n+1} = I_{n+1}$	$A_1 - B_1\Delta_1/2 + C_1(\Delta_1/2)^2 =$ $A_N + B_N\Delta_N/2 + C_N(\Delta_N/2)^2 =$
7 Three-term sinusoidal with point matched weights and extrapolated currents	Hallen (YEH and MEI [4.27]) Pocklington (GEE et al. [4.33])	$A_n + B_n\sin(ks'_n) + C_n\sin(ks'_n)$	$A_n - B_n\sin(k\Delta_{n-1,n}) + C_n\cos(k\Delta_{n-1,n})$ $= I_{n-1}$ $A_n + C_n = I_n$ $A_n + B_n\sin(k\Delta_{n,n+1}) + C_n\cos(k\Delta_{n,n+1})$ $= I_{n+1}$	$A_1 - B_1\sin(k\Delta_1/2)$ $+ C_1\cos(k\Delta_1/2) = 0$ $A_N + B_N\sin(k\Delta_N/2)$ $+ C_N\cos(k\Delta_N/2) = 0$
8 Three-term sinusoidal with point matched weights and junctioncurrent amplitude and slope matching	Pocklington	$A_n + B_n\sin(ks'_n) + C_n\cos(ks'_n)$	$A_n + B_n\sin(k\Delta_n/2) + C_n\cos(k\Delta_n/2)$ $= A_{n+1} - B_{n+1}\sin(k\Delta_{n+1}/2)$ $+ C_n\cos(k\Delta_{n+1}/2)$ $A_n + C_n = I_n$ $B_n\cos(k\Delta_n/2) - C_n\sin(k\Delta_n/2) = B_{n+1}$ $\cos(k\Delta_{n+1}/2) + C_{n+1}\sin(k\Delta_{n+1}/2)$	$A_1 - B_1\sin(k\Delta_1/2)$ $+ C_1\cos(k\Delta_1/2) = 0$ $A_N + B_N\sin(k\Delta_N/2)$ $C_N\cos(k\Delta_N/2) = 0$
9 Three-term sinusoidal with point match weights and junction matched currents only	Pocklington	$A_n^{(+)} + B_n^{(+)}\sin(ks'_n)$ $C_n^{(+)}\cos(ks'_n) + A_n^{(-)}$ $+ B_n^{(-)}\sin(ks'_n) + C_n^{(-)}\cos(ks'_n)$	$A_n^{(+)} + B_n^{(+)}\sin(k\Delta_n/2) + C_n^{(+)}\cos(k\Delta_n/2)$ $= I_{n,n+1}$ $A_n^{(+)} - B_n^{(+)}\sin(k\Delta_n/2) + C_n^{(+)}\cos(k\Delta_n/2) = 0$ $B_n^{(+)}\cos(k\Delta_n/2) + C_n^{(+)}\sin(k\Delta_n/2) = 0$ $A_n^{(-)} - B_n^{(-)}\sin(k\Delta_n/2) + C_n^{(-)}\cos(k\Delta_n/2)$ $= I_{n-1,n}$ $A_n^{(-)} + B_n^{(-)}\sin(k\Delta_n/2)$ $+ C_n^{(-)}\cos(k\Delta_n/2) = 0$ $B_n^{(-)}\cos(k\Delta_n/2) - C_n^{(-)}\sin(k\Delta_n/2) = 0$	 $C_1^{(+)} = A_1^{(-)} = B_1^{(-)} = C_1^{(-)} =$ $C_N^{(-)} = A_N^{(+)} = B_N^{(+)} = C_N^{(+)} =$ $A_1^{(+)} + B_1^{(+)}\sin(k\Delta_1/2) = I_{1,2}$ $A_1^{(+)} - B_1^{(+)}\sin(k\Delta_1/2) = 0$ $A_N^{(-)} - B_N^{(-)}\sin(k\Delta_N/2)$ $= I_{N-1,N}$ $A_N^{(-)} + B_N^{(-)}\sin(k\Delta_N/2) = 0$

ss otherwise Indicated $s'_n = s - s_n$ $s_{n+} = s - s_n + \Delta_n/2$ $\Delta_{N+1} = \Delta_0 \equiv 0$ for open ended wires

$= 1, \ldots, N.$ $\Delta_{n,n+1} \equiv (\Delta_n + \Delta_{n+1})/2$ $s_{n-} = s_n - s + \Delta_n/2$ $\Delta_{N+1} \equiv \Delta_1$, $\Delta_0 \equiv \Delta_N$ for loops

s function in terms of unknowns rior segment	End segment (if different)	Weight function Interior segment End segment (if differ.)	Number of unknowns Open ended wires	Loops	Comments
	—	$\delta(s - s_m)$	N	N	Least rapidly convergent of methods considered here (THEILE [4.9]). Has discontinuous charge.
	—	$\delta(s - s_m)$	N	N	The finite difference form for the charge implies a linear current variation between I_n and I_{n+1} for purposes of defining the derivative $I'_{n,n+1}$ which applies between s_n and s_{n+1}. Has discontinuous current.
$_{+1}\sin(ks_{n+}) + I_{n-1,n}\sin(ks_{n-})]/\cdot\sin(k\Delta_n)$	$I_{1,2}\sin(ks_{1+})/\sin(k\Delta_1)$ $I_{N-1,N}\sin(ks_{N-})/\sin(k\Delta_N)$	$[\Delta_m\delta(s - s_m)$ $+ \Delta_{m+1}\delta(s - s_{m+1})]/$ $(\Delta_m + \Delta_{m+1})$	$N-1$	N	Not useful except for cases where $\Delta \sim a$ (TAYLOR [4.29]). Has discontinuous charge.
$_{+1}s_{n+} + I_{n-1,n}s_{n-})/\Delta_n$	$I_{1,2}s_{1+}/\Delta_n$ $I_{N-1,N}s_{N-}/\Delta_N$	1 s_{1+}/Δ_1 s_{N-}/Δ_N (for end segment)	$N-1$	N	The integrations involved over the current and field are each handled with two integrand samples per segment by CHAO and STRAIT [4.4].
$_{+1}\sin(ks_{n+}) + I_{n-1,n}\sin(ks_{n-})]/\sin(k\Delta_n)$	$I_{1,2}\sin(ks_{1+})/\sin(k\Delta_1)$ $I_{N-1,N}\sin(ks_{N-})\sin(k\Delta_N)$	$2\sin(k\Delta_m/2) \times$ $\cos(ks'_m)/\sin(k\Delta_m)$ $\sin(ks_{1+})/\sin(k\Delta_1/2)$ $\sin(ks_{N-})/\sin(k\Delta_N/2)$ (for end segment)	$N-1$	N	This form has the advantage of an analytically integrable basis, as well as a more accurate representation than a pulse or linear basis. Has discontinuous charge.
$_{n,n+1}(1 - s'^2_n) + s'_n(\Delta^2_{n,n+1} - 1)]\, I_{n-1}$ $[s'_n(\Delta^2_{n-1,n} - \Delta^2_{n,n+1}) + s'^2_n(\Delta_{n-1,n} + \Delta_{n,n+1})$ $\Delta_{n,n+1}\Delta^2_{n-1,n} - \Delta_{n-1,n}\Delta^2_{n,n+1}]\, I_n$ $[\Delta_{n-1,n}(1 - s'^2_n) + s'_n(1 - \Delta^2_{n-1,n})\, I_{n+1}\}\, D^{-1}$ $\Delta_{n,n+1}(1 - \Delta^2_{n-1,n}) + \Delta_{n-1,n}(1 - \Delta^2_{n,n+1})$	For segment 1 use with $I_0 = \Delta_0 \equiv 0$ For segment N use with $I_{N+1} = \Delta_{N+1} \equiv 0$	$\delta(s - s_m)$	N	N	This form can be used in the same way as the three-term sinusoidal bases (Methods 7–9). This example corresponds to method 7 and has a discontinuous current.
$k\Delta_{n,n+1})[1 - \cos(ks'_n)] + \sin(ks'_n)$ $\mathrm{os}(k\Delta_{n,n+1}) - 1])\, I_{n-1} + (\sin(ks'_n)[\cos(k\Delta_{n-1,n})$ $\cos(k\Delta_{n,n+1})] + \cos(ks'_n)[\sin(k\Delta_{n,n+1})$ $\sin(k\Delta_{n,n-1})] - \sin(k\Delta_{n,n+1})\cos(k\Delta_{n-1,n})$ $\sin(k\Delta_{n-1,n})\cos(k\Delta_{n,n+1}))\, I_n$ $(\sin(k\Delta_{n-1,n})[1 - \cos(k\Delta s'_n)] + \sin(ks'_n)$ $- \cos(k\Delta_{n-1,n})])\, I_{n+1}\}\, D^{-1}$ $\sin(k\Delta_{n,n+1})[1 - \cos(k\Delta_{n-1,n})]$ $\sin(k\Delta_{n-1,n})\,[1 - \cos(k\Delta_{n,n+1})]$	For segment 1 use with $I_0 = \Delta_0 \equiv 0$ For segment N use with $I_{N+1} = \Delta_{N+1} \equiv 0$	$\delta(s - s_m)$	N	N	Has discontinuous current.
ıot be written for an individual segment. Expansion th segment involves current samples on all N segments.		$\delta(s - s_m)$	N	N	Has discontinuous charge derivative, but continuous current and charge.
$_{,n} + I_{n,n+1})/[2\sin^2(k\Delta_n/2)]$ $\sin(ks'_n)(I_{n-1,n} - I_{n,n+1})/[2\sin(k\Delta_n/2)]$ $\cos(k\Delta_n/2)\cos(ks'_n)(I_{n-1,n} + I_{n,n+1})/$ $\sin^2(k\Delta_n/2)]$ $[(I_{n-1,n} + I_{n,n+1}) - \cos(ks_{n+})\, I_{n,n+1}$ $\cos(ks_{n-})\, I_{n-1,n}]/2\sin^2(k\Delta_n/2)$	$I_{1,2}/2 + \sin(ks'_1)/$ $[2\sin(k\Delta_1/2)]$ $I_{N-1,N}/2 - \sin(ks'_N)/$ $[2\sin(k\Delta_N/2)]$	$[\Delta_m\delta(s - s_m)$ $+ \Delta_{m+1}\delta(s - s_{m+1})]/$ $(\Delta_m + \Delta_{m+1})$	$N-1$	N	Has discontinuous charge.

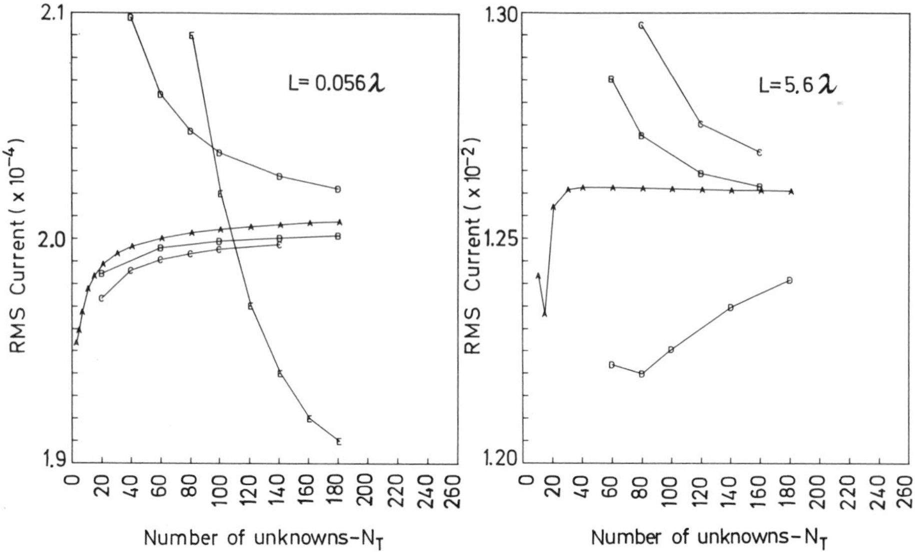

Fig. 4.1. Convergence rate as a function of the number of current samples for several current basis functions (see Table 4.2 for a description of the current basis functions). (A: Method 7, B: Method 5, C: Method 4, D: Method 2, E: Method 1)

much computer time is associated with the overall computation. For all methods outlined, the impedance matrix fill time will generally exceed the subsequent matrix solution time for $N \lesssim 200$. This point is discussed further in Section 4.9.

A number of comparisons have been made concerning the convergence rates of various of the methods summarized in the table. (See, for example, NEUREUTHER et al. [4.8], THIELE [4.9], MILLER and DEADRICK [4.10], LOGAN [4.11], MILLER et al. [4.12].) The results of such studies have demonstrated among other things that:

1) sinusoidal bases are superior for structures whose maximum dimension exceeds a few wave lengths,

2) the three-term basis used with point matching (e.g., Method 8) is nearly equivalent to the piecewise sinusoidal basis used with Galerkin's method (Method 5),

3) pulse bases can generate acceptable results if care is exercised in treating the corresponding charge, i.e., their use in the vector-scalar potential integral equation with due account taken of the charge term via implied interpolation (Method 2), but not in the Pocklington integral equation (Method 1),

4) a piecewise sinusoidal basis does not in general produce satisfactory results with point matching (Method 3),

5) the convergence rate of a numerical solution can be meaningfully expressed in terms of a sample density per unit wavelength only for longer structures, (total length $L \gtrsim 3\lambda$), while for shorter structures, it is generally the number of samples required to adequately define the geometry which is the determining factor.

Some results from one of these studies are shown in Fig. 4.1 [4.12]. Presented here are results for I_{RMS} pertaining to the current induced on a straight wire for a normally incident plane wave, where specifically

$$I_{\mathrm{RMS}}(N_{\mathrm{T}}) = \left[1/N_{\mathrm{T}} \sum_{i=1}^{N_{\mathrm{T}}} |I_i|^2 \right]^{1/2}$$

with N_{T} the number of current samples for the test case, and I_i the corresponding current values at the N_{T} sample points. The data of Fig. 4.1 demonstrates the superiority of the sinusoidal bases function types for the longer wire ($L \approx 5.6\lambda$), while also showing the more nearly equal convergence rates obtained for the shorter wire ($L \approx 0.05\lambda$). Similar findings were obtained for plane wave scattering from an L-wire and two crossed wires having the same total lengths.

4.3. Structural Segmentation and Boundary Condition Matching

The two general aspects which must be considered when employing a moment method, the current basis function and weight function used, were discussed in a general way above. As mentioned there, these two questions are not unrelated since it is often convenient to refer the current expansion and the boundary-condition match points to the same set of coordinates on the structure. In employing subsectional collocation for example, we actually collocate the field match points and the current sample points. Even where the boundary condition matching is more general as in the approximate Galerkin's method employed by Chao and Strait [4.4] in connection with the piecewise linear current expansion or the piecewise sinusoidal approach of Richmond [4.6], the field match points are referred in a fixed way to the current segment. Thus, when we speak of developing a numerical model for a wire structure we must not only consider the impact of approximating the current distribution in an acceptable way, but of matching the boundary conditions of the total field along the structure as well. In the subsections which follow we consider first the convergence rate of the numerical solution as a function of the number of segments for a variety of wire structures. We next discuss the impact of structural segmentation on boundary matching in the source region of antennas, and conclude with the effect of impedance loading on boundary condition matching.

4.3.1. Structural Segmentation

The development of a suitable numerical model for a given wire structure is of vital importance in obtaining efficient and accurate numerical results. For simpler structures such as the straight wire, circular ring, etc., there is not too much latitude available to the user. What is primarily required is a determination of the sample density necessary per unit wave length, or minimum number of samples required, at which point the development of a numerical model is straight-forward and rather obvious. When, on the other hand, the structure being considered is more complicated, a Rhombic antenna for example, or conical spiral

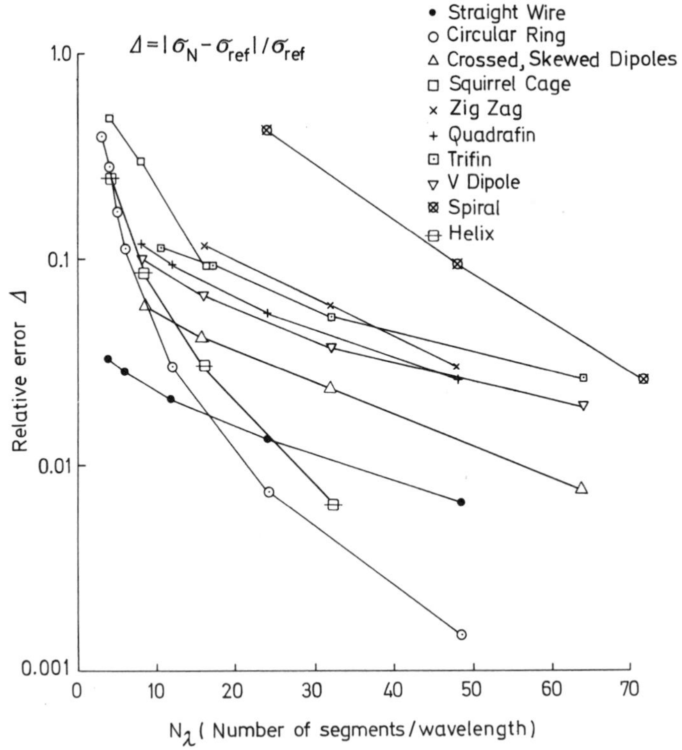

Fig. 4.2. Relative error vs segments wavelength for various structures. (After MILLER et. al. [4.13])

and other more general kinds of antennas, the parameters of the numerical model to be used are no longer so clearly defined.

Some rule-of-thumb guidelines have been developed for a selection of representative structures encountered in practical applications. Such studies, one of which is summarized in Fig. 4.2 due to MILLER et al. [4.13], have shown that, on the order of 6–20 samples per wave length are required to achieve an accuracy of approximately 10% or better in computed radar cross-section results, antenna input impedance, etc. [1]. In addition, it has been found that where practicable the adjacent segments on a structure ought not to change too abruptly in length, although the influence of this factor is related to the particular problem involved [2]. In actual applications this is quite frequently not achievable, for example, the feed region of a complicated antenna may be very small relative to the antenna size. Clearly one would not like to use small segments over the entire structure since this increases the computer time quite drastically.

What is required in that case is some compromise, possibly using smaller segments near the feed region and letting the segments become longer as the distance from the feed region increases.

4.3.2. Boundary Condition Matching

The use of equal length segments is intuitively acceptable and apparently mathematically desirable too for many problems. By matching the boundary conditions at segment centers (using subsectional collocation) or some equivalent approach, the use of equal length segments insures that the boundary match points are equally spaced along the structure as well. However, there is no necessity for collocating match points and segment centers. If the condition on the field match points is relaxed to allow them to be located at evenly-spaced points along the structure while varying segment lengths are used to accommodate more rapid current variations in the vicinity of junctions and sources, possibly a more efficient numerical procedure might be developed. This kind of approach has not been investigated to any significant extent apparently. Care must be excercised in considering this more general approach

[1] As mentioned above, the sampling density required for structures $\sim \lambda$ in size are related more to the requirements for defining the structure's geometry than to the wavelength.

[2] It should be noted that the method 7) based on the three-term current basis and delta function weights appears more sensitive to having unequal segments in certain circumstances than some of the other commonly used procedures. The section which follows derives from a study to identify possible deficiencies in source description and junction treatments which might affect accuracy achieved in boundary condition matching.

incidentally, since field match points located near discontinuities in the current or its derivative might lead to other numerical problems.

Besides the requirement that the total tangential field be zero at points on the structure away from the source, it is vitally important in performing antenna calculations to properly introduce the exciting field in the calculation. This is generally associated with the one (or more) segments on the structure which act as the source region. Such source segments thus serve to induce currents over the entire structure. Before computation of the antenna input admittance or impedance is possible, however, the voltage associated with the source must be established. If all equal length segments are used for a structure modeled with a point matching approach, it is generally accurate to consider the applied voltage on a given source segment to equal the product of this segment length and the negative of the electric field value at its center. This is an approximation though, which may not always be satisfied and which may furthermore be extremely sensitive to the structure segmentation.

As an example of the potential problems associated with structure segmentation and boundary condition matching we present in Table 4.3 some results obtained from a systematic variation of a dipole-antenna model which nominally employed 21 equal length segments with the center segment serving as the source. The tangential electric field on the source segment is taken as the negative voltage divided by the segment length, i.e., $E = -V/\delta$. For convenience we let V equal 1 V.

Three variations from the nominal model were studied. In the first the center, or feed, segment was systematically shortened relative to the remaining 20 segments on the antenna so that the ratio R of the feed segment length to the length of the remaining 20 equal length segments varied from 1 to 1/64 with the results shown in A of Table 4.3. There we see that the input admittance defined by $Y = I_{\text{feed}}/(-E\delta)$ rapidly changes as a function of R, thus generating some skepticism as to the accuracy of the numerical results. Similar results are obtained when current and charge matching at segment junctions is employed. While the dependence of the input susceptance of an antenna upon the feed region geometry is well known [4.14] it is also equally known that the conductance should be relatively independent of the feed region geometry, contrary to the present results. Therefore we conclude that something has been made invalid as a result of varying the feed segment length relative to the uniform segment lengths on the remainder of the antenna.

A part of the explanation may be due to the fact that the actual antenna input voltage is no longer really one volt. Recall that a tangential field is used as an exciting source, with the subsequent driving voltage derived by multiplying the field strength by the length of the excited segment. Since we have no control over what happens to the electric

Table 4.3. Segment length variation for a center-fed dipole. Shown for comparison are results obtained with junction current amplitude and charge matching, and the bicone source model

A. Single variable feed segment of length δ

$Y = I_{\text{feed}}/(-E^{\text{I}}\delta)^{\text{a}}$ $\quad L = \lambda/2 = (N-1)\Delta + \delta^{\text{b}}$

$\Omega = 15^{\text{e}}$

$N = 21^{\text{b}}$

Tangential E source model			Bicone source model[c]
	Current extrapolation Method 7	Current and charge matching at segment ends Method 8	Current and charge matching at segment ends Method 8
$R = \delta/\Delta$	Y [Millimhos]	Y [Millimhos]	Y [Millimhos]
1/1	9.55 − j 5.16	9.38 − j 5.26	8.69 − j 4.77
1/2	11.74 − j 6.31	10.04 − j 5.59	8.60 − j 4.71
1/4	14.13 − j 7.57	11.34 − j 6.28	8.63 − j 4.72
1/8	15.99 − j 8.55	13.75 − j 7.59	8.81 − j 4.81
1/16	17.98 − j 9.60	18.34 − j 10.10	9.00 − j 4.90
1/32	23.53 − j 12.56	27.65 − j 15.22	9.08 − j 4.80
1/64	41.36 − j 22.08	47.78 − j 26.81	6.52 − j 3.61 [d]

B. Three variable length segments at center $L = \lambda/2 = (N-3)\Delta + 3\delta$

C. Two variable length segments ($n = 5, 17$) $L = \lambda/2 = (N-2)\Delta + 2\delta$

$R = \delta/\Delta$	B. Y [Millimhos]	C. Y [Millimhos]
1	9.55 − j 5.16	9.55 − j 5.16
1/2	9.57 − j 5.13	9.57 − j 5.16
1/4	9.57 − j 5.16	9.57 − j 5.16
1/8	10.13 − j 5.39	9.57 − j 5.16
1/16	11.45 − j 6.08	9.57 − j 5.16
1/32	14.06 − j 7.45	9.57 − j 5.16
1/64	16.19 − j 8.57	9.57 − j 5.16

[a] Applies to all results except bicone source model (to be defined in Table 4.7).

[b] Applies to all results except bicone source model where $N = 20$ and the two center segments were both of length δ.

[c] After Poggio et al. [4.36].

[d] Demonstrates onset of thin-wire approximation becoming invalid ($2a/\Delta = 0.31$, see Section 4.5).

[e] Ω denotes the expansion parameter of Hallén ($\Omega = 2 \ln$ [length/radius]).

field between the field match points on the antenna we might suspect that by changing the feed segment relative to the others, the field variation between the segments has changed in a way which has caused the driving voltage defined as above to be incorrect.

One possible way to maintain more control over the feed region field variation is to use segments on either side of the feed segment which are equal in length to it, so that the field becomes zero at one segment on either side of the applied electric field itself. The results of this parameter variation are shown in Section B of Table 4.3 where the center three segments on the antenna now have the length δ and the remaining 18 segments the length Δ. The results are again displayed as a function of $R = \delta/\Delta$. In this case we find that as R becomes smaller, the input admittance remains relatively stable until R reaches a value of 1/16 or so. Beyond this point, there is again found the variation of input admittance previously demonstrated in Section A of the table.

Finally in Section C of Table 4.3 we examine the influence of placing variable length segments away from the feed region. The positions chosen are five segments from each end of the dipole. Again, we denote the lengths of the shortened segments by δ and the remaining segment lengths by Δ with $R = \delta/\Delta$. The input admittance in this case is relatively insensitive to the variation of R. This is significant in that it shows that unequal length segments located near, or in, the feed region have a much more profound influence on the calculated input admittance of an antenna than segments of unequal length located farther away.

It has been mentioned that the reason for this variation in the input admittance as a function of the source segment length may be that the applied field no longer results in a one volt source across the center of the antenna. The possibility that this is the case may be studied by integrating the electric field along the antenna in the vicinity of the source segment. We therefore present results in Table 4.4 corresponding to those of Table 4.3 but where the input admittance is now defined to be the ratio of the feed point current divided by the integrated electric field, i.e., $Y = I_{\text{feed}}/-\int E\,dl$. Section A again corresponds to the case of a single variable length segment and Section B to three variable length segments. In contrast to the previous case where significant variations of input admittance were found as R became smaller, we now find that the input admittance remains relatively insensitive to R. The integrated voltage value used to define the input admittance changes in such a way as to compensate for the input current variation, and thereby yields a nearly constant input admittance value. It should be noted that the integration range extends over approximately the center third of the antenna, and is terminated when the last segment integrated changes the voltage value by less than one percent.

This is an interesting and potentially useful result. While in practical cases we would naturally want to avoid the additional expense involved in integrating the electric field along the antenna to obtain a valid input admittance, at the same time knowledge that this might be a method

Table 4.4. Segment variation for center-fed dipole using current extrapolation (Method 7)
$\Omega = 15$
$N = 21$
$Y = I_{\text{feed}}/(-\int E(s')\,ds')$

A. Single variable feed segment of length δ

$R = \delta/\Delta$	$V = -\int E(s')\,ds'$ [V]	Y [Millimhos]
1	1.000 + j 0.015	9.46 − j 5.31
1/2	1.227 + j 0.019	9.48 − j 5.29
1/4	1.476 + j 0.022	9.49 − j 5.17
1/8	1.670 + j 0.027	9.49 − j 5.27
1/16	1.876 + j 0.030	9.50 − j 5.27
1/32	2.456 + j 0.039	9.50 − j 5.27
1/64	4.321 + j 0.069	9.49 − j 5.26

B. Three variable length segments at center

$R = \delta/\Delta$	$V = -\int E(s')\,ds'$ [V]	Y [Millimhos]
1	1.000 + j 0.015	9.46 − j 5.31
1/2	1.001 + j 0.021	9.45 − j 5.32
1/4	1.006 + j 0.022	9.49 − j 5.33
1/8	1.052 + j 0.023	9.51 − j 5.33
1/16	1.187 + j 0.026	9.53 − j 5.33
1/32	1.454 + j 0.031	9.55 − j 5.33
1/64	1.675 + j 0.037	9.55 − j 5.33

whereby the input admittance could be obtained is worthwhile. Furthermore, we know that calculations for radiation pattern, etc., which may not depend on knowing the input power or input admittance of the antenna could also be valid. It might therefore be concluded that for reasonably simple antennas useful input impedance and admittance data can be obtained by first deriving the voltage from an integral of the electric field, a procedure essentially equivalent to the classical emf method. Actual field variations along the antenna are shown later in Section 4.7.

It is also worthwhile to determine the influence of segment length variations on the calculated input admittance for the antenna where the structure has a multiple junction. An antenna consisting of a straight center section and two outward pointing V loads having a 60° included angle symetrically located on each end of the center section was studied. In the nominal configuration this antenna has a total of 39 equal length segments; 3 on the center and 9 on each of the four arms. Two segment length variations were investigated for this particular structure. In

Table 4.5. Segment length variation for center-fed dipole with V end loads

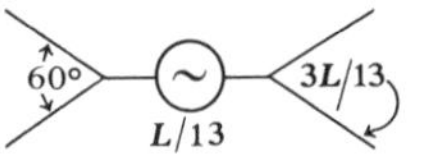

$a/\lambda = 2.798 \times 10^{-4}$
$L/\lambda = 0.75 = (N - M)\,\Delta + M\,\delta$
$Y = I_{\text{feed}}/(-E^{\text{I}}\delta)$

A. M equal length segments on center section[a]

$R = \delta/\Delta$	N	M	Y [Millimhos]
3/3	39	3	19.7 +j 12.2
3/5	41	5	23.1 +j 10.0
3/7	43	7	27.7 +j 4.2
3/9	45	9	28.5 −j 5.9
3/11	47	11	22.2 −j 14.36

B. Divisions of the center segment into M equal-length parts

$R = \delta/\Delta$	N	M	Y [Millimhos]
1	39	1	19.7 +j 12.2
1/3	41	3	19.6 +j 12.4
1/5	43	5	19.6 +j 12.4
1/7	45	7	19.7 +j 12.5
1/9	47	9	19.8 +j 12.5
1/11	49	11	19.8 +j 12.5

[a] Results obtained by employing Curtis's [4.15] charge treatment for the junction together with the three-term current vary only slightly over the δ/Δ range shown (see Table 4.6). The results of this and other methods are presented by MILLER et al. [4.12].

the first, results of which are summarized in Table 4.5, Section A, the segment lengths on the center portion of the antenna were systematically shortened by using 5, 7, 9, etc., segments in place of the original 3. Corresponding input admittance results are indicated also in Table 4.5, Section A, where the variation of the input admittance obtained from an assumed applied voltage of one volt is seen to be significant as the number of center segments and hence the ratio R is varied. In Section B are the summarized results wherein only the center segment was shortened by replacing it with 3, 5, 7, etc., segments and exciting the center of those segments. Results obtained here exhibit hardly any variation in input admittance, indicating perhaps that the unequal length segments at the multiple junction in the former case play a significant role in the admittance variation.

This study was repeated by defining the input admittance as a ratio of the feed current to the integral of the applied electric field in the vicinity of the source region. Results obtained are summarized in

Table 4.6. Segment length variation for center-fed dipole with V end loads

$$a/\lambda = 2.798 \times 10^{-4}$$
$$L/\lambda = 0.75 = (N - M)\,\Delta + M\,\delta$$
$$Y = I_{\text{feed}}/(-\int E(s')\,ds')$$

A. M equal length segments on center section

$R = \delta/\Delta$	N	M	$-\int E(s')\,ds'$	Y [Millimhos]
3/3	39	3	1.25 − j 0.26	13.13 + j 12.46
3/5	41	5	1.30 − j 0.48	13.39 + j 12.35
3/7	43	7	1.27 − j 0.89	13.06 + j 12.51
3/9	45	9	0.92 − j 1.33	12.02 + j 12.47
3/11	47	11	0.34 − j 1.44	13.18 + j 12.74

B. Division of the center segment into M equal-length parts

$R = \delta/\Delta$	N	M	$-\int E(s')\,ds'$	Y [Millimhos]
1/1	39	1	1.25 − j 0.26	13.13 + j 12.46
1/3	41	3	1.25 − j 0.25	13.21 + j 12.58
1/5	43	5	1.25 − j 0.25	13.08 + j 12.54
1/7	45	7	1.25 − j 0.26	12.94 + j 12.33
1/9	47	9	1.25 − j 0.26	13.16 + j 12.67
1/11	49	11	1.25 − j 0.26	13.05 + j 12.54

Table 4.6. In a manner similar to the linear dipole example discussed above we find that the applied voltage shown in Table 4.6 does vary systematically as the feed region geometry is changed, and in a way which leads to the relatively stable value of input admittance as found for the linear dipole. These results indicate that integrating the electric field to define the antenna input admittance is apparently a useful procedure for general geometries. Note that the necessity for doing so is evidently related to the multiple junction treatment used, as the admittance variation found here does not occur in results obtained by CURTIS [4.15], STRAIT [4.16], and using Richmond's approach [4.12], nor in time domain calculation which employ a junction treatment analogous to that used here [4.17] and where $V = -E\delta$ is used rather than an integral definition of V.

A summary of some source models which have been used for wire antennas analysis is given in Table 4.7. Also see KING [4.18] for a discussion of the source problems. Generally speaking we conclude that models which produce sharply localized tangential fields (Methods 2 and 3) are preferable.

Table 4.7. Summary of some antenna source models

Method	Reference	Applied with (see Table 4.2)	Source specified via	Admittance calculation	Comments
1 Tangential E-field on segment p	MILLER [4.14], NEUREUTHER et al. [4.8]	Any point matching approach; 1, 2, 3, 6, 7, 8, 9.	$E(s=s_p)=-V/\Delta_p$	$I(s=s_p)/[-E(s=s_p)\Delta_p]$	Requires well-behaved E-field. Has advantage that solution matrix is source independent.
2 Current slope discontinuity between segments p and p+1	ANDREASEN and HARRIS [4.28]	Any approach where I' can be specified.	$[I'(s_{p+1}-\Delta_{p+1}/2+\varepsilon)$ $-I'(s_p+\Delta_p/2-\varepsilon)]$ $=jkV/Z_0=(I'_{p+1}-I'_p)$ $\lim \varepsilon\to 0$	$I(s=s_p+\Delta_p/2)/$ $[z_0(I'_{p+1}-I'_p)/jk]$	Z_0 is impedance computed for bicone antenna having the radius and length of the source segments. [4.29]. Results in a sharply peaked and localized E-field which resembles a delta-function.
3 Delta-function E-field at junction of segments p and p+1	RICHMOND [4.6] CHAO and STRAIT [4.4]	Galerkin's method, methods 4 and 5.	$E(s)=-V\,\delta(s-s_p-\Delta_p/2)$	$I(s=s_p+\Delta_p/2)/V$	Cannot be used with point matching, but is similar to current slope discontinuity.
4 Current source on segment p	THIELE [4.30]	Any method	$I(s=s_p)$	$I(s=s_p)/[-\int E(s)\,ds]$	Has disadvantages of requiring numerical field integration for admittance determination. Has apparently not been widely used.
5 Magnetic frill at s_p on segment p	TSAI [4.31]	Any method	$E(s)=1/[2\ln(b/a)]$ $\cdot(e^{-jkR_a}/R_a-e^{-jkR_b}/R_b)$ $R_x=\sqrt{(s-s_p)^2+x^2}$ a, b = frill inner and outer radii.	$I(s=s_p)/[-\int E(s)\,ds]$	May require numerical integration. Of limited use with point matching if fields are ill-behaved between match points.

4.3.3. Impedance Loading

Not all wire structures that are of practical concern can be considered to be perfectly conducting. There are two ways in which the effects of finite wire conductivity may become important. The first and most obvious occurs when the wire does indeed have a finite conductivity or where the skin depth is large enough compared to the wire diameter that the assumption of vanishing electric field along the wire is no longer strictly valid. This particular situation can be characterized as equivalent to a distributed load along the wire. The second occurs due to impedance loads located at discrete points along the wire. These might be included to modify the structure's resonance characteristics or to provide matched operation, e.g., the load on the end of a two-wire transmission line. In the latter case then, the effect of the lumped impedance load is confined to a particular localized point(s) on the wire structure. An experimental validation of the former has been given by CASSIDY and FEYNBERG [4.19]. The latter case has been validated by computations for a two-wire transmission line having varying load impedance values [4.3].

Either of these two cases is suited for treatment via the perfectly conducting wire integral Eq. (4.1) presented in Section 4.1 by a suitable modification of the boundary condition matching along the wire. In the context of the point matching solution to the integral equation we generalize the boundary conditions by allowing for the additional effect of the voltage drop at each match point in terms of the particular value of impedance located there. A distributed impedance load will in general then be modeled using a non-zero load impedance for each segment of the wire structure. In the case of lumped loading, on the other hand, only a few segments will have an associated impedance.

An important practical application of impedance loading is that of modeling insulators. In many low frequency antenna problems insulators are employed to electrically isolate the antenna from its support structures, to make the guy wire system electrically small to minimize re-radiation, etc. It is of interest in this application to know the effective voltage drop which appears across these insulators for design optimization and to permit realistic maintenance schedules to be set up.

Various alternatives are available to us to determine the actual insulator voltage drop. We might, for example, use an impedance load of high value at the point of the insulator location and determine the voltage drop across the insulator from the product of the current times the insulator impedance value along the loaded segment. Or perhaps we might use the value of electric field at the center of the segment and multiply this by the segment length to approximate the insulator voltage drop. Alternatively we might, as in the approach used for an open circuit transmission line [4.3], leave a physical gap in the structure at the insulator's position and then subsequently compute the electric field along the location of the missing insulator to obtain a voltage drop. The voltage drop along the insulator could also be approximated by the product of the field value at the center of the gap times the insulator length.

There are obvious advantages to being able to employ the lumped impedance load insulator model. Perhaps most important is that the effect of varying the insulator impedance values on the structures characteristics can be readily assessed to determine the effect of a deteriorating insulator on the structure's electromagnetic performance. Such information would be especially valuable for determining maintenance schedules. It is, however, not clear that a value for the insulator voltage drop obtained in this way will be entirely valid because in our model the current will only vanish (or become small) at the point of the load itself, but in general will be non-zero over the rest of the load, whereas the real insulator would have no appreciable current flow anywhere on it. The potential validity of the lumped impedance load as an insulator model thus obviously depends upon both the current basis and weight functions employed. A physical gap may, on the other hand, represent a more realistic model for the insulator, but evaluation of the insulator voltage drop will require additional field computation(s).

In order to assess the relative advantages of modeling insulators with physical gaps versus lumped impedance loads we have performed the following computations. The first pertains to the two-wire transmission line model already mentioned while the second set applies to the case of a linear dipole antenna.

The results of the calculations are presented in Table 4.8. Section A is for a 600 Ω transmission line model, for which analytical results for the voltage across the terminating load resistor are available. It can be seen that we obtain an essentially correct result whether we integrate the tangential electric field across the loaded segment, or multiply the current evaluated at the center point location of the loaded segment times the load resistence value.

Section B of Table 4.8 is for a linear half-wavelength dipole nominally having 21 equal length segments with two symmetrically located insulators centered 1/8 wavelength from each end. The insulators were modeled in three ways: 1) with lumped resistive loading on segments 5, 6, and 16, 17; 2) with the loaded segments replaced by a physical gap, and 3) with a physical gap one segment long located such that the gap end points coincide with the current match points of the loaded segments in Model 1.

We see from the results presented in Section B of Table 4.8 that we can obtain an approximate measure of the integral of the tangential

Table 4.8. Insulator models

A. 600 Ω 2-wire transmission line excited by a matched 1 V source

Termination	$E(s')\,ds'$	$I_L R_L$	Analytic
Matched load ($R_L = 600\ \Omega$)	0.46 V	0.48 V	0.50
Open circuit ($R_L = 10^{10}\ \Omega$)	0.97 V	1.03 V	1.00
Short circuit ($R_L = 0\ \Omega$)	0.002 V	0.00 V	0.00

B. Dipole antenna, 1 W input power, 21 segments

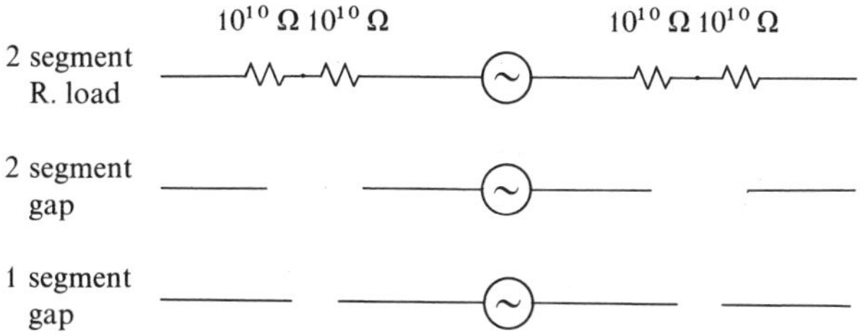

	2 segments loaded	2-segment gap	1-segment gap
$\int E(s')\,ds'$	: 321 V	113 V	105 V
$I_L R_L$	: 192 V	—	—
$E^*_{T\text{Mid}} L_g$	: 192 V	37 V	36.4 V
Z_{in}	: 12.1 − j 678	10.0 − j 761	12.8 − j 663

electric field across a gap insulator by simply computing the IR drop for the resistively loaded segments. A comparison of the results obtained by the loaded segment method versus the 1 and 2 segment physical gap insulator model shows the impedance load voltage drop to be substantially higher. The input impedance obtained for the one-segment gap is seen to agree more closely with the loaded segment case, apparently because the points where the current goes to zero in the loaded case are effectively the segment centers, which coincide with the physical ends of the one-segment gap. This is caused by the three-term current extrapolation procedure and might be partially alleviated by using junction matching (Method 8 of Table 4.2).

4.4. Multiple Junction Treatment

Application of the methods considered above in Section 4.2 was discussed for simply connected structures having no multiple junctions (three or more wires meeting at a common point). In Table 4.9 which follows, we

Table 4.9. Some multiple junction treatments

	Method number of basis and weight functions (see Table 4.2)	Reference	Multiple junction treatment	Comments
1			None required	
2	2	CURTIS [4.15]. See SAYRE [4.32] for similar treatment	$I'_{n,n+1} \rightarrow \left(\sum_{i=1}^{M} I_{n+i} - I_n \right) 2A_n/(A_T \Delta_n)$	Derived on basis of assuming the charge on junction wire n is distributed according to the ratio of the area of wire $n(A_n)$ to the total area of the $M+1$ junction wires (A_T).
3	3, 4, 5, 9	CHAO and STRAIT [4.4] RICHMOND [4.6]	$I_{n,n+1} \rightarrow \sum_{i=1}^{M} I_{J,n+i}$	Enforces Kirkhoff's law at the junction.
4	3, 4, 5, 9	CHAO and STRAIT [4.4] RICHMOND [4.6]	$I_{n,n+1} \rightarrow I_{J,n}$ $I_{J,n+1} = I_{J,n} + \tilde{I}_{J,n+1}$ $I_{J,n+i} = \tilde{I}_{J,n+i-1} + \tilde{I}_{J,n+i} \quad i=2,\dots,M$ $\tilde{I}_{J,n+M} = 0$	Equivalent to above, but written in terms of overlapped basis, where the "$\sim$" denotes the overlapped segment currents.

5	6, 7	GEE et al. [4.33]	$I_{n+1} \to \sum_{i=1}^{M} I_{n+i}$ $\Delta_{n,n+1} \to \left[\Delta_n + (1/M) \sum_{i=1}^{M} \Delta_{n+i}\right]/2$	Extrapolates to a "pseudo" segment having the average length and total current of the M-connected segments.
6	6, 7	MILLER and DEADRICK [4.3]	$I_{n+1} \to I_n(1 - A_n/A_T) + A_n/A_T \sum_{i=1}^{M} I_{n+i}$ $\Delta_{n,n+1} \to \Delta_n/2$	An adaptation of the CURTIS [4.16] charge treatment to the three-term basis.
7	8	ANDREASON and HARRIS [4.28]	The bicone-antenna junction model	This method is considerably more complex than those discussed here, but is evidently the only approach which takes the local junction geometry into account. Results obtained by AREN [4.34] for the V-dipole test case previously considered (for M equal-length segments on the center) were more variables as M was increased than Methods 2, 4, 5, 7, the latter using the Curtis charge junction treatment (see Methods 2, 4, 5, 7 of Table 4.2)

summarize some of the multiple junction treatments which have been employed with these methods. For simplicity, we consider a segment n connected at its positive end to M other wires whose reference directions are out of the junction. The results are shown as modifications to the plus end current relations previously given for segment n and the two-wire junction in Table 4.2. We denote the current on a given segment i, $i=n+1,\ldots,n+M$, at its multiple junction (negative) end as $I_{J,i}$. Note that the various multiple junction treatments used with a three-term basis function are essentially interchangeable, i.e., a given multiple junction approach might be used with any of the methods which employ a three-term basis. On a comparative basis, and using information presently available, the junction treatments can be classified in decreasing order of validity as 3 (or 4), 2, 6, 5, and 1, omitting method 8 because of insufficient data.

4.5. The Thin-Wire Approximation

There are two potential problems which one may encounter using the thin-wire approximation. One of these is its unsuitability for application to wires more than a small fraction of a wavelength in diameter. This is caused by the fact that in deriving the thin-wire kernel the azimuthal variation of the current around the wire is ignored as well as the azimuthal variation of the kernel in the integral equation itself. Both are replaced by the mean values.

The other potential problem which arises from using the thin-wire approximation is based on the numerical method used for solving the integral equation. The problem involves the use of segments whose lengths are shorter than several wire diameters. In this case, the nature of the integral equation kernel may produce non-physical current oscillations near junctions and source regions which are numerically generated.

In order to address the question of segment size relative to the wire diameter, we have determined the input admittance and current distribution on a rather fat ($\Omega = 8$) half-wave antenna as a function of the number of segments. Results obtained are summarized in Table 4.10 and Figs. 4.3 and 4.4. We present in Table 4.10 the computed input admittance for this antenna as a function of the ratio of the wire radius a to the segment length Δ. The input admittance results which are obtained exhibit a conductance which is insensitive to this ratio, but the susceptance may be seen to vary dramatically and become obviously invalid as this ratio is decreased. The reason for this occurrence is dramatized by the current distributions of Fig. 4.3 where the real current is seen to exhibit

Table 4.10. Dipole short segment admittance

$L/\lambda = 0.5$
$a/\lambda = 0.00916$
$\Omega = 8$
$Y_{\mathrm{King}} \sim 10.06 - \mathrm{j}\,3.6$

N	a/Δ	$Y = I_{\mathrm{feed}}/(-E\,\delta)$
21	0.38	$7.7 - \mathrm{j}\ \ 3.3$
81	1.48	$7.5 + \mathrm{j}\ \ 0.97$
181	3.32	$7.2 + \mathrm{j}\,343.2$

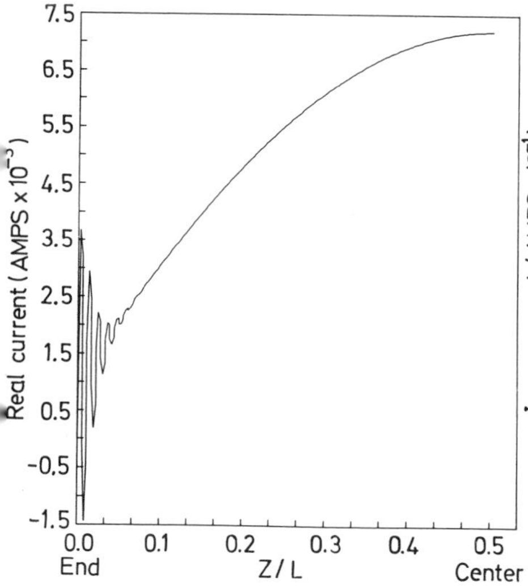

Fig. 4.3

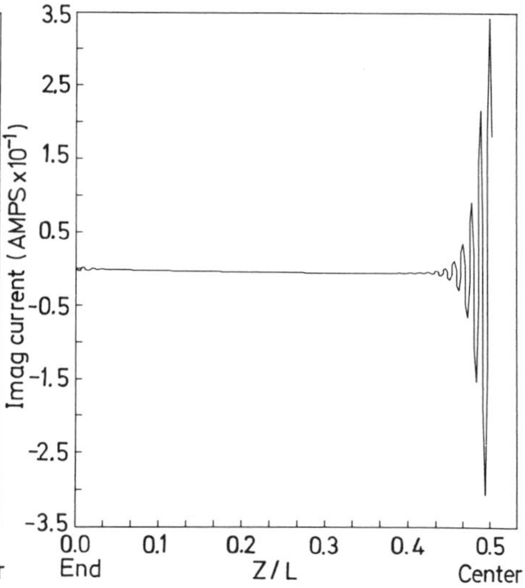

Fig. 4.4

Fig. 4.3. Current on a half-wavelength dipole modeled with short segments ($\Omega = 8$, 181 segments, frequency: 1 MHz)

Fig. 4.4. Current on a half-wavelength dipole modeled with short segments ($\Omega = 8$, 181 segments, frequency: 1 MHz)

an oscillation at the end of the antenna. The imaginary current also exhibits an oscillation near the end of the antenna but, in addition, one much more pronounced near the feed region itself. These results are for the extreme case of 181 segments and an a/Δ ratio of 3.32. The imaginary current oscillation near the feed region is seen to have values on an order of 100 times larger than the mean current away from the feed region. Obviously any attempt to define an input susceptance on the basis of a current that varies so rapidly near the feed region is questionable at best. However, if one smoothly extrapolates back to the feed region the antenna current behavior away from the oscillatory portion, one does obtain in all cases a reasonably stable and consistent input susceptance. Therefore while the current distribution itself contains obviously invalid behavior, it is still possible to derive what seem to be valid admittance results. Furthermore, the antenna radiation pattern is essentially unaffected by the oscillatory current since the current oscillation naturally tends to cancel out and thus disappear from the far-field calculation itself.

Thus, while one should be alert to the potential problems involved in using what are sometimes termed pancake shaped segments, i.e., segments shorter than a wire diameter or so in length, it is still feasible to obtain useful calculated data if one is careful to define input admittance in terms of an extrapolated current.

4.6. Matrix Factorization Roundoff Error

In order to demonstrate in a controlled way the result of matrix roundoff error on the final numerical result a series of computations have been performed for various types of wire structures. Results of these calculations are summarized in Fig. 4.5. There we plot the error in calculated input impedance for several cases as a function of the number of bits in the coefficients of the impedance matrix. Two structures are shown. One is a straight wire having 7, 21, and 121 segments. The other involves two different Loran antennas, the Sectionalized Loran Transmitting (SLT) antenna modeled with 237 segments, and the Top Loaded Monopole (TLM) antenna with 240 segments. The initial impedance matrix is computed in the usual fashion. A modified matrix is then obtained by systematically truncating each matrix (mantissa) word entry and retaining the number of bits indicated on the plot. The truncated matrix is then factored to obtain the antenna current distribution and input impedance in the usual way. The error in the input impedance is finally defined relative to the value obtained for the non-truncated case [4.3].

As expected, the solution accuracy increases as the number of bits in the matrix word size is increased. The relationship is essentially a logarithmic one, i.e., the numerical accuracy is exponentially related to the number of bits in the impedance matrix coefficients. Furthermore, as is also expected, the accuracy decreases as the number of unknowns or linear system size increases.

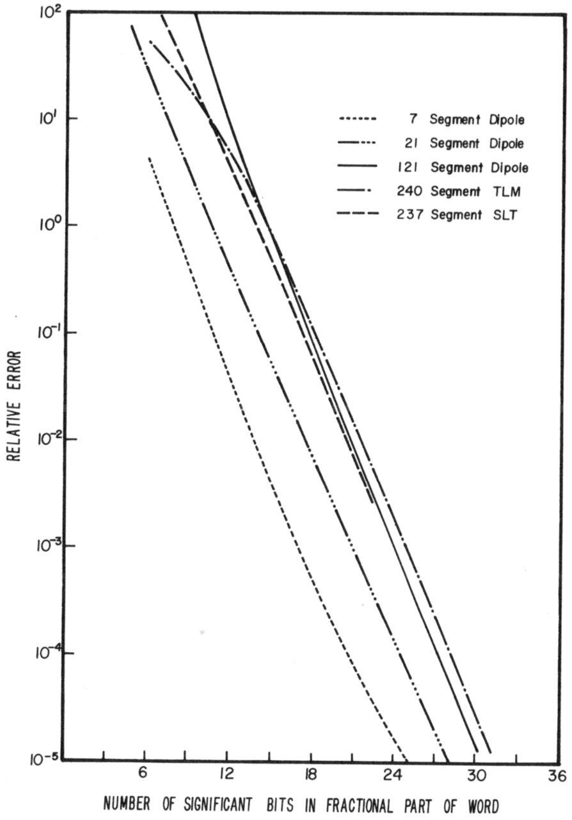

Fig. 4.5. Impedance error vs computer word size

4.7. Near-Field Behavior

The basis for all numerical modeling via integral equations in electromagnetics is the computation of fields due to current and charge sources. Presumably, then, if one obtains a valid solution as measured by the accuracy of input admittance, radar cross-section, etc., it must have required the calculation of numerically valid fields in originating the impedance matrix itself. Yet the demonstrated accuracy of input admittance, radar cross-section etc., is no guarantee that the overall solution can be accepted as being physically realistic or correct. As a matter of fact, various aspects of the calculated results may be obviously invalid, but without a negative impact on the overall usefulness of the calculation. One example of this is demonstrated above in connection with the oscillatory nature of the imaginary part of the antenna current near the source region for cases where the thin-wire kernel becomes invalid (see Section 4.5 preceeding).

The near fields around antennas are of significant interest with respect to corona discharge assessment, determining insulator voltage requirements, EMP vulnerability assessment, etc. Consequently, we have a legitimate interest in their behavior. However, inaccurate artifacts may be introduced into the field calculation by pecularities associated with the current basis-function expansion.

Let us first consider how the radial field E_ϱ along a straight wire depends upon the current distribution. Since the magnitude of E_ϱ is proportional to the surface charge density, we have

$$E_\varrho \propto d[I(z)]/dz .$$

The various basis functions discussed above lead to currents which may have discontinuities in amplitude, slope, etc., at segment junctions. Each of these can thus produce a distinctive variation in E_ϱ at the discontinuity. The current amplitude discontinuity results in a delta-function behavior in E_ϱ, while the slope discontinuity produces a step in E_ϱ, for example. The longitudinal electric field component E_z is also affected by such characteristics of the numerical solution. Since E_z involves an integral of the current multiplied by a kernel having a second derivative with respect to z of the Green's function, (4.1), it will exhibit a behavior reflecting the second derivative of the current. Consequently, the current amplitude discontinuity leads to a doublet behavior in E_z, the slope discontinuity to a delta-function behavior, and a charge-derivative discontinuity to a step behavior in E_z. A current distribution having a continuous second derivative would thus be required for a smooth field distribution to be obtained along the wire.

Examples of the tangential field along one half of a linear half-wave antenna excited by a one-volt source are presented on a log scale in Fig. 4.6. These results were obtained from: 1) a pulse current basis and point matching, 2) the piece-wise sinusoidal basis with Galerkin's method, 3) the three-term basis with point matching and the current extrapolated to adjacent segment centers, and 4) same as 3) but with slope and amplitude matching of the current at the junctions[3]. We observe that, as expected, the pulse basis leads to the most ill-behaved tangential field variation while the three-term basis with junction matched slope (charge) and amplitude is best behaved. In all but the Galerkin's method the fields exhibit numerical zeroes at the segment centers. Note that the input impedance values obtained from these

[3] Methods 1, 5, 7, and 8, respectively, from Table 4.2.

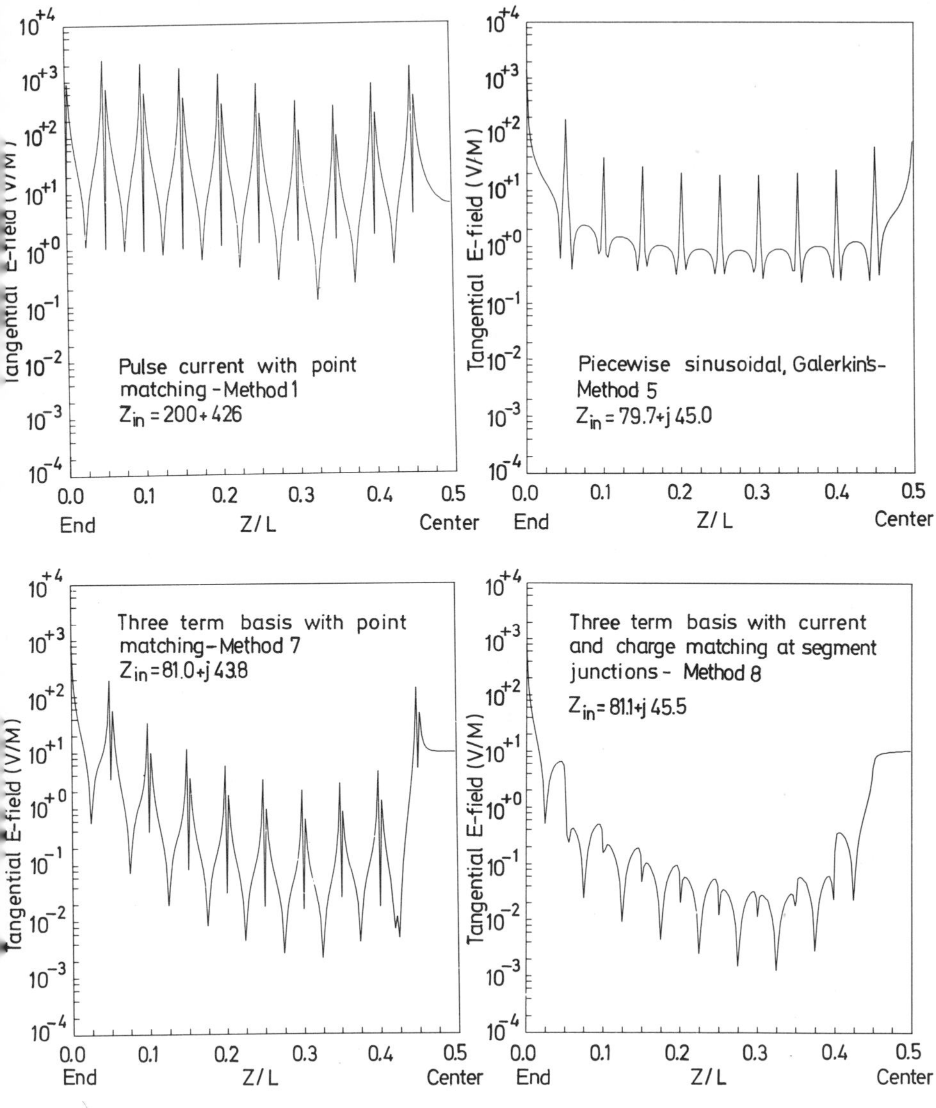

Fig. 4.6. Tangential fields along half of a half-wavelength dipole for various current basis functions. Methods referenced are found in Table 4.2.

examples are reasonably close in spite of their substantially different tangential fields. Finally, note that the source fields (near the right-hand side of each graph) are quite different, reflecting the influence of the various source and current treatments employed.

4.8. Wire-Grid Modeling

Wire-grid meshes find many uses in application where the effect of a solid conducting surface is required but the weight and/or wind resistance of the latter must be avoided. They may be used, for example, to fabricate radar antenna reflectors, and as shields to screen sensitive equipment from stray fields. Their successful substitution for the solid surface depends upon the fact that as the mesh size becomes smaller relative to the shortest wavelength of concern, the mesh supports a surface current distribution which approaches that on the continuous surface. This phenomena occurs because of the transverse nature of electromagnetic fields, and does not hold for an acoustic field, which is longitudinal, for example.

Exploitation of wire-grids as substitutes for solid surfaces need not be confined to actual practice, but can also be advantageous for the analytical study of certain problem types. Many problems of practical interest, for example radar cross-section studies, antenna analysis and EMP interactions, involve hybrid geometries which have features of both solid surfaces and wire-like structures. As such, these geometries are not well suited for treatment via the magnetic field integral equation (MFIE) (although this equation is apparently superior to the electric-field type for structures consisting only of solid surfaces). Furthermore, the MFIE is not suited for the analysis of shell-like solid surfaces, such as thin plates and spherical shell sections. The thin-wire approximation to the electric field integral equation (EFIE) (4.1), however, offers a formulation which can in principle be used for treating all these geometries by modeling the solid surface parts of such structures with wire-grids. By extension, of course, wire-grids might then be used as well for the computer modeling of solid surface structures which might normally be treated via the MFIE.

As a matter of fact, wire-grid models have already been quite widely used for a variety of problems and geometries. RICHMOND [4.21] was apparently the first to report on the application of the thin-wire EFIE to the analysis of wire-grid models for circular disks and spheres. He demonstrated satisfactory agreement between his wire grid results and independent analytical or experimental back-scatter cross-section data presented as a function of frequency.

Subsequently, other wire-grid results have been described by TANNER and ANDREASON [4.24], MILLER and MAXUM [4.22], MILLER and MORTON [4.20], THIELE [4.21], DIAZ [4.23], and POGGIO and MILLER [4.2]. These studies included the modeling of airplanes and helicopters (TANNER and ANDREASON, MILLER and MAXUM, and DIAZ), a flat-backed cone (THIELE), and flat plates (MILLER and MORTON). Representative wire-grid model

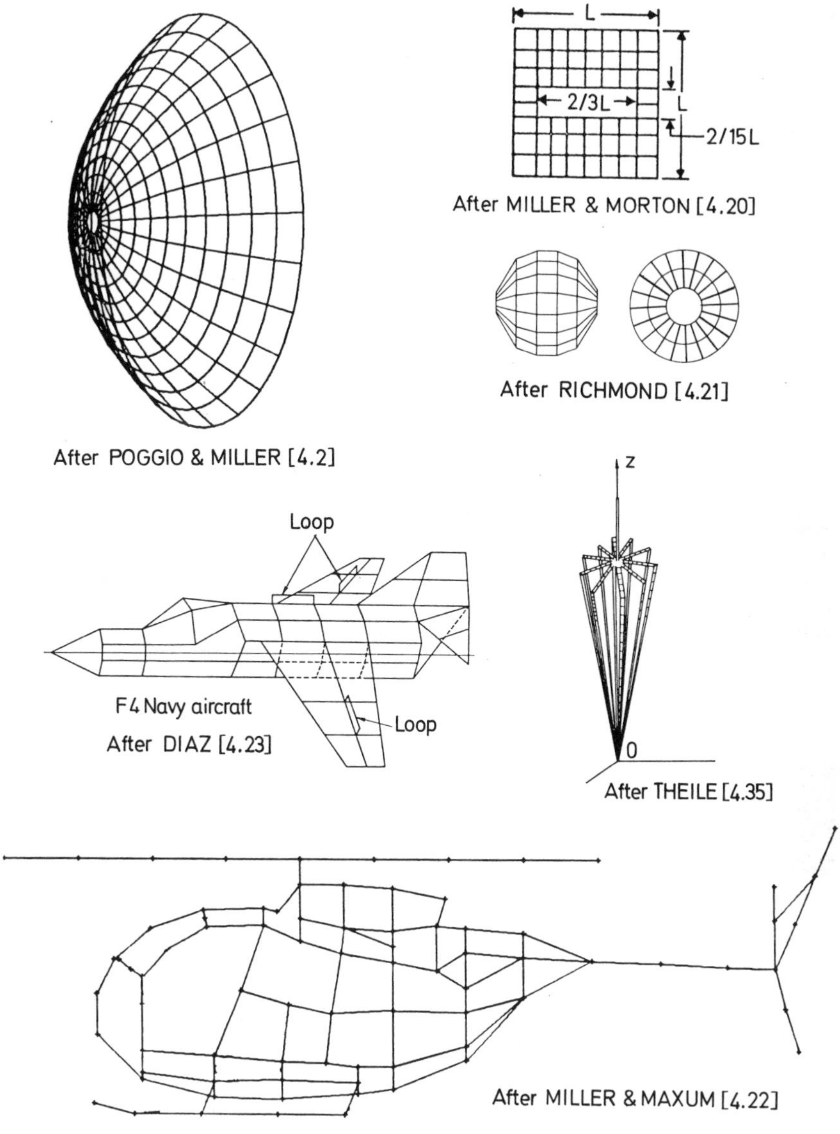

Fig. 4.7. Representative wire grid model structures

structures are depicted in Fig. 4.7. In all cases, there was generally found to be reasonable agreement between theory and experiment for a far-field quantity such as the radar cross-section or antenna radiation pattern with the agreement relatively independent of the excitation (e.g., flat plate scattering was equally good for edge-on and broadside incidence).

These studies, while illustrating the applicability of wire-grid meshes as models for solid surfaces in terms of their far-field electromagnetic behavior are not entirely convincing as to the use of wire-grid models to determine near field quantities, such as charge density and surface current distributions. Preliminary studies in this regard to compare the results obtained with analytical models like spheres and circular disks with their wire-grid counterparts are not yet conclusive. Such comparisons should do much to more clearly define the areas of application and limitations of wire-grid models.

One particular test case which has been investigated and which does shed some light on the potential problems inherent in using wire-grid models concerns a simple linear dipole antenna. The dipole antenna considered had an Ω value of 6 and is thus not well suited to treatment via the thin-wire integral equation because of the dipole's thickness. Therefore, BURKE and SELDEN [4.25] have developed a simple wire model of this particular antenna by representing it with four equispaced, parallel wires whose centers lie on the dipole's circumference. In order to determine the sensitivity of the results to the end of the antenna, two sets of calculations were performed; one with the thin-wires open-ended and the other with the wires cross-connected at the ends of the dipole.

The input impedance was obtained as a function of frequency for the two models, results of which for the cross-connected case are presented in Fig. 4.8. These impedance values were obtained for two different kinds of excitation, symmetric and asymmetric. In the former, the center segments on all four wires were excited with the same electric field, corresponding to a one-volt source. For the asymmetric excitation, only one of the four wires was excited, again with the field corresponding to a one-volt source. The input admittance in each case was defined to be the sum of the currents on the center segments of the four wires. Both models yielded the value for input impedance shown by the solid curve on the figures. This result is in fairly good agreement with that presented by KING [4.18].

A significant difference is seen, however, in the individual wire currents for the asymmetric excitation. The variation with frequency of the individual currents on the three wires, as indicated on the figure, differs markedly from their sum. As a matter of fact, the individual wire currents can become as large as two orders of magnitude greater than the resultant current for the four center segments. This situation evidently arises because the two pairs of wires effectively act as transmission lines whose currents are thus oppositely directed and of nearly the same magnitude. In order that the sum be that for the symmetrically excited

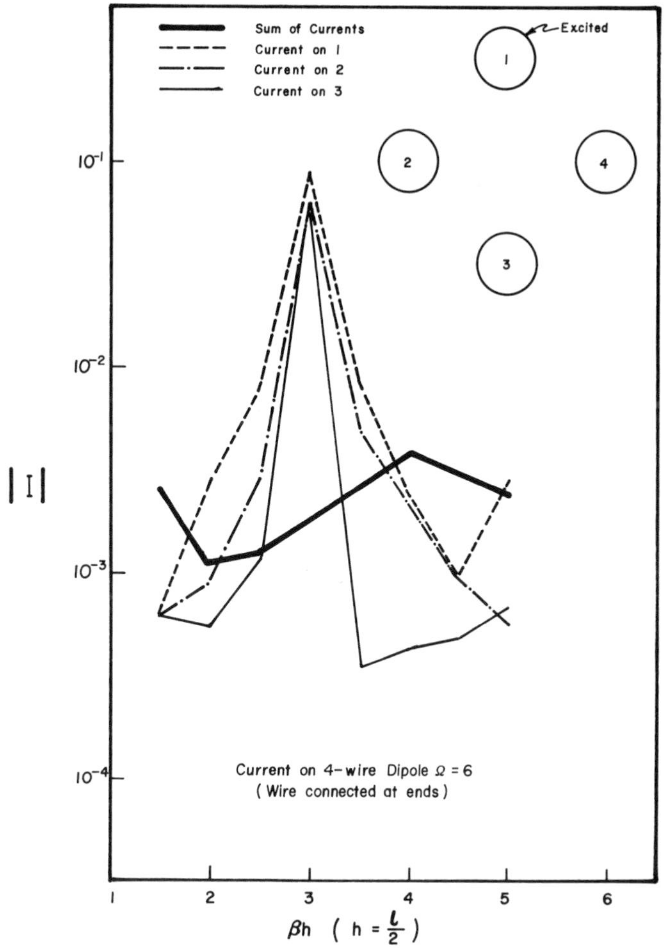

Fig. 4.8. Current on 4-wire dipole $\Omega = 6$ (wire connected at ends). (After BURKE and SELDEN [4.25])

configuration, the individual wire currents can consequently become very large. This result indicates that asymmetric excitation of the wire-grid model for a solid surface may produce individual wire currents which are significantly different from those which would occur on a solid conducting surface itself. While in this case the composite effect did not exhibit a significant error, such a circumstance can, of course, not always be guaranteed. A similar behavior has been encountered on other kinds of wire-grid models for solid surfaces, e.g., large circulating currents are found to occur on the interior wire loops of wire-grids [4.25].

4.9. Computer Time

In prior discussion we have not dealt in any specific way with the actual computer time requirements, or what is equivalent, the expense of performing a given calculation. Results such as those presented above in Section 4.4 are useful to demonstrate the convergence rate of a numerical solution as a function of the number of segments and to develop application guidelines. Such results do not, however, indicate in general what method might offer the greatest efficiency to obtain a given accuracy in any particular application. The reason for this is that while the number of samples required per wavelength to achieve a given accuracy is one indicator of numerical efficiency, that alone does not provide a complete basis for comparison with other methods. In addition to the convergence rate itself, we must be concerned with the computer time required to compute the matrix, since it is this factor together with the convergence rate which determines the overall efficiency of a particular method.

The computer time associated with the moment method or matrix solution of an integral equation for a wire or surface structure can be approximated from the following equation

$$t \cong AN^2 + BN^3 + CN^2 N_i + DNN_iN_a ,$$

where N is the order of the linear system or the number of unknowns, N_i is the number of incident fields or source configurations, and N_a is the number of observation points in the far field. The corresponding Coefficients A, B, C, and D represent the computer time associated with computing the impedance matrix, factoring or inverting it, computing the currents from the specified incident field and computing the far field, respectively. Obviously, a comparison of two different methods for solving a given problem is incomplete without considering both the number of samples, N, required to obtain a given accuracy from each and the timing coefficients associated with their use, i.e., we must compare t_1/t_2 rather than N_1/N_2 alone. Note that the Coefficients A, B, C, and D are thus both algorithm and computer dependent.

It is therefore necessary when comparing alternate methods to have available the timing coefficients associated with their application referred to a common computer. Unfortunately such data is not easy to obtain. The various programs have generally been developed for use on different computers, have not been optimized and have not been used to analyze the same set of problems. We have, however, been able to directly compare the running times on the CDC-7600 for several programs.

Table 4.11. Computer timing coefficients (CDC-7600)

Time $A N^2 + B N^3$

Numerical method		Solution time [sec]		
	A	50 segs	100 segs	200 segs
Piecewise sinusoidal, Galerkin's [4.6]	3.5×10^{-4}	0.99	4.4	22.00
Piecewise linear, Galerkin's [4.4]	2.6×10^{-4}	0.77	3.5	18.00
Pulse with point matching [4.15]	3.0×10^{-4}	0.87	3.9	20.00
Three-term sinusoidal with point matching [4.22]	4.2×10^{-4}	1.20	5.1	24.00

For all methods we use $B \approx 9.4 \times 10^{-7}$

These results are summarized in Table 4.11. The actual matrix fill-time coefficient is shown for each method.

Only a representative value for the matrix-solution coefficient is shown rather than those of each program, since both inversion and factorization have been used in the various programs, and this factor is essentially independent of the integral equation and numerical method used. The current computation and far field evaluation coefficients are not included since they are usually of minor import.

For convenience, we have given the matrix fill-time coefficient A for each of these methods and, in addition, the representative computer time required to solve a problem with $N = 50$, 100, and 200 segments. Generally speaking, there is not a substantial amount of difference between their required computer times. It must be recognized that these computer programs are probably not optimized and that furthermore the details of the computation and consequent relative running times may vary from structure to structure. These data, however, do provide some indication of the computer times involved in employing such programs. Note also that symmetry can be exploited in various ways to reduce both computer storage and running times [4.10].

4.10. Observations and Conclusion

The preceeding discussion was intended to not only demonstrate some of the capabilities inherent in wire integral equation analysis, but also to emphasize some of the pitfalls and problems which can be only too easily encountered in its use. Such numerical anomalies as negative input resistance for an isolated antenna; a current oscillation related to the sampling interval rather than the wavelength; or the divergence of a numerical result with increasing sample density, are obvious enough to alert even the inexperienced user to question a given result, although their rectification may not be equally obvious. The more subtle the problem, however, the more difficult identification of its occurrence and estimation of its impact on the validity of the calculated data. Consequently as a necessary adjunct to the sequence of operations involved in the overall modeling procedure; 1) theoretical formulation, 2) mathematical manipulation; and 3) numerical computation, is 4) validation of the calculated result. It is probably safe to say that this last step may ultimately absorb as much time in the course of using a given program as the former three.

If it is concluded that validation is a key element in permitting a numerical method to be reliably and confidently used, we must then ask how this can best be realized. There are two obvious approaches: 1) experimental-comparison of the calculated results with measured data to estimate the physical modeling error ε_p, and 2) numerical-comparison with independent theoretical results and/or internal numerical consistency checks to evaluate the numerical modeling error ε_N.

Experimental validation is generally the most convincing, but measurement has its own difficulties which quite often frustrate its effectiveness in this role. While comparison of experimental and computed data in an absolute sense is probably preferable, much can be done on a relative basis by, for example, comparing the difference between calculations for two numerical models with that obtained from measurements for two corresponding experimental models. Also on a relative basis, two experimental models might be measured. The first should correspond as closely as possible to the numerical model to provide a check on the *numerical* validity of the calculation. The second should resemble the real problems as closely as possible to provide a check on the physical validity of the numerical model, which for all but the simplest problems will incorporate some degree of approximation, or which due to the formulation may completely alter the problem description. As an example of the former, we might use circular cylinders and flat plates to model an aircraft, while in the latter instance, we might instead use a wire-grid model.

As an addition or alternative to experimental validation we might choose a numerical route. Comparisons with independent theoretical results can involve steps quite similar to those discussed above in connection with experimental validation. We might, for example, compare wire-grid results with analytical solutions for disks, spheroids, etc. The other area is that of internal consistency checks. Some possibilities are:

1) Determine whether the bistatic scattered fields and mutual admittances satisfy reciprocity.

2) Evaluate the degree to which energy is conserved.
3) Insert the numerical solution into the original linear system.
4) Increase sampling density.
5) Plot current and charge distributions.
6) Examine the tangential fields to see the degree of boundary condition accuracy between match points.

Of the above, 1) and 2) are necessary but not sufficient conditions for solution validity. The third provides a check on possible matrix roundoff, while 4) should provide an indication of the numerical convergence. Item 5) may indicate faulty results through obviously non-physical current and charge distributions. The sixth could probably be the most definitive, but at the same time, most expensive to use. In addition, as has been shown above, the fields along the structure can exhibit obvious errors while the solution retains good accuracy.

An additional check we have not mentioned is physical intuition, a sometimes useful resource, but one unfortunately of uneven quality. It all too frequently happens that suspicious-looking numerical (or for that matter experimental) data can after a time be explained on plausible physical grounds, followed then by discovery of the computational (experimental) error which was really the culprit.

In spite of the various kinds of checks to which the validation of computer produced results can be subjected, perhaps the best check of all is careful attention to detail in computer usage, both in the development of an algorithm and its subsequent application. Certainly the computer can provide no better output than the instructions and input which are given it. It is extremely important to keep an open mind if the computer is to be most effectively exploited, both to question results which appear wrong, and to seek an explanation for results which conflict with our preconceptions.

Acknowledgements

Support for the work presented here was primarily provided by the Department of Transportation, United States Coast Guard. The authors are especially grateful to Mr. WALTER O. HENRY of the Coast Guard for his interest in this study and for making it possible. They also appreciate the cooperation of: G.J. BURKE and E.S. SELDEN of MB Associates, Prof. B.J. STRAIT of Syracuse University, and Mr. W. CURTIS of Boeing Aircraft Co., who provided data and helpful information. The work was performed under the auspices of the U.S. Atomic Energy Comission.

References

4.1. R. F. HARRINGTON: *Field Computation by Moment Methods* (The Macmillan Company, New York, 1968).
4.2. A. J. POGGIO, E. K. MILLER: In *Computer Techniques for Electromagnetics*, ed. by R. MITTRA (Pergamon Press, New York, 1973), Chapt. IV.
4.3. E. K. MILLER, F. J. DEADRICK: "Some Computational Aspects of Thin-Wire Modeling"; Tech. Report UCRL 74818, Lawrence Livermore Laboratory (1973).
4.4. H. H. CHAO, B. J. STRAIT: IEEE Trans. Antennas Propagation AP-**19**, 701 (1971).
4.5. K. K. MEI: IEEE Trans. Antennas Propagation AP-**13**, 374 (1968).
4.6. J. H. RICHMOND: "Computer Analysis of Three-Dimensional Wire Antennas"; Tech. Report 2708-4, Ohio State University, Electro-Science Lab. (1969).
4.7. J. H. RICHMOND: Proc. IEEE **53**, 796 (1965).
4.8. A. R. NEUREUTHER, B. D. FULLER, G. D. HAKKE, G. HOHMANN: "A Comparison of Numerical Methods for Thin Wire Antennas"; Presented at Fall URSI meeting, University of California, Berkeley, Calif. (1968).
4.9. G. A. THIELE: In *Computer Techniques for Electromagnetics*, ed. by R. MITTRA (Pergamon Press, New York, 1973), Chapt. II.
4.10. E. K. MILLER, F. J. DEADRICK: "Computer Modeling of LORAN Antennas"; Tech. Report UCRL 51464, Lawrence Livermore Laboratory (October 1973).
4.11. J. C. LOGAN: "A Comparison of Techniques for Treating Radiation and Scattering by Wire Configurations and Junctions"; Tech. Report TR-73-10, Syracuse University (1973).
4.12. E. K. MILLER, R. M. BEVENSEE, A. J. POGGIO, R. W. ADAMS, F. J. DEADRICK, J. A. LANDT: "An Evaluation of Computer Program Using Integral Equations for the Electromagnetic Analysis of Thin-Wire Structures"; Tech. Report UCID 75566, Lawrence Livermore Laboratory (March 1974).
4.13. E. K. MILLER, G. J. BURKE, E. S. SELDEN: IEEE Trans. Antennas Propagation AP-**19**, 534 (1971).
4.14. E. K. MILLER: Radio Sci. **2**, 1431 (1967).
4.15. W. L. CURTIS: Private communication (1972).
4.16. B. J. STRAIT: Private communication (1973).
4.17. J. A. LANDT: Private communication (1973).
4.18. R. W. P. KING: *The Theory of Linear Antennas* (Harvard University Press, Cambridge, Mass., 1956).
4.19. E. A. CASSIDY, J. FEYNBERG: IRE Trans. Antennas Propagation AP-**8**, 1 (1968).
4.20. E. K. MILLER, J. B. MORTON: IEEE Trans. Antennas Propagation AP-**18**, 290 (1970).
4.21. J. H. RICHMOND: IEEE Trans. Antennas Propagation AP-**14**, 782 (1966).
4.22. E. K. MILLER, B. J. MAXUM: "Mathematical Modeling of Aircraft Antennas and Supporting Structures"; Final Report ECOM Contract ADDB07-68-C-0456, Report No. ECOM-0456-1 (1970).
4.23. M. DIAZ: "Computer Techniques for Electromagnetics and Antennas"; Short Corse Notes Chapter 2, University of Illinois (28 Sept.–1 Oct. 1970), Fig. 2–47.
4.24. R. L. TANNER, M. G. ANDREASEN: IEEE Spectrum **4**, 53 (1967).
4.25. G. J. BURKE, E. S. SELDEN: Private communication (1972).
4.26. C. D. TAYLOR: Private communication (1973).
4.27. Y. S. YEH, K. K. MEI: IEEE Trans. Antennas Propagation AP-**15**, 634 (1967).
4.28. M. G. ANDREASEN, F. G. HARRIS, JR.: "Analysis of Wire Antennas of Arbitrary Configuration by Precise Theoretical Numerical Techniques"; Tech. Report ECOM 0631-F, Contract DAAB07-67-C-0631, Granger Associates, Palo Alto, Calif. (1968).
4.29. S. A. SCHELKUNOFF: *Advanced Antenna Theory* (J. Wiley & Sons, New York, 1952).

4.30. G. A. Thiele: Private communication (1970).
4.31. L. L. Tsai: "Near and Far Fields of a Magnetic Field Current"; Digest of the URSI Spring Meeting (1970).
4.32. E. P. Sayre: IEEE Trans. Antennas Propagation AP-**21**, 216 (1973).
4.33. S. Gee, E. K. Miller, A. J. Poggio, E. S. Selden, G. J. Burke: "Computer Techniques for Electromagnetic and Radiation Analyses"; Internat. EMC Symposium Record, Philadelphia, Pa. (1971).
4.34. V. A. Arens: Private communication (1973).
4.35. G. A. Thiele, M. Diaz: *Radiation of a Monopole Antenna on the Base of a Conical Structure* (Proc. Conf. Environmental Effects on Antenna Performance, J. R. Wait, Editor, Environmental Science Services Administration, Boulder, Colo., 1969).
4.36. A. J. Poggio, R. W. Adams, E. K. Miller: "A Study of Antenna Source Models"; Tech. Report UCRL 51693, Lawrence Livermore Laboratory (October 1974).

5. Stability and Convergence of Moment Method Solutions

R. MITTRA and C. A. KLEIN

With 23 Figures

The material presented in this chapter is divided into two broad areas. In the first part we discuss the problem of stability of integral equation solutions derived via matrix methods. We introduce a quantity called the "condition number" which is useful not only for identifying the unstable regimes in the solution procedure but also for evaluating different techniques that might be employed for extracting stable solutions from ill-conditioned equations.

The second part of this chapter deals with the question of convergence of solution of matrix equations. Several different tests for convergence are examined in considerable detail for the thin-wire antenna problem and some relevant guidelines based on these studies are presented.

5.1. Stability

5.1.1. Use of Condition Numbers

Most problems in electromagnetics that can be formulated as an integral equation are now routinely solved by using the method of moments to convert the continuous integral equation into a discrete matrix form which can be easily solved on modern computers. For example, problems expressed as a Fredholm equation of the first kind

$$\int_a^b K(x, y)\, f(x)\, dx = g(y) \tag{5.1}$$

in which one wishes to determine some function $f(x)$ from its integral transform $g(y)$, become approximated by determining some vector f given the vector g; i.e.

$$Af = g\,. \tag{5.2}$$

For inverse or remote sensing problems, such as the determination of the profile of an inhomogeneous medium, $g(y)$ is derived from experimentally gathered data. In these cases one typically finds that (5.2) is unstable, i.e., small errors in g produce large errors in the calculated value of f. For other problems, such as finding the current distribution on antennas, $g(y)$ is analytically known, but it is necessary to approximate the kernel in determining the matrix A. For example, the elements of A may be the result of numerical integration. In both cases, therefore, it would be useful to be able to estimate the error in f in terms of errors in either g or A.

It will be shown that a quantity to be identified as a matrix condition number can detect situations such as resonances which cause ill-conditioned results [5.1]. In other problems the condition number can show the correct choice of some parameter which would at first sight appear to be arbitrary. When an ill-conditioned situation is encountered, the condition number can be used to evaluate various methods for changing the formulation to obtain an acceptable answer.

Matrix condition numbers can be easily developed from simple inequalities involving vector and matrix norms. Since a wide variety of definitions of matrix norms for the general case of complex elements exists in the literature, it is worthwhile to be specific about the precise definition of all norms to be used. Toward this end we first define A^{H} and x^{H} to indicate the transpose conjugate of matrix A and vector x, respectively. The familiar Euclidean vector norm is then defined by

$$\|x\|_2 = \sqrt{x^{\mathrm{H}} x} = \sqrt{|x_1|^2 + |x_2|^2 + \cdots + |x_n|^2}\,, \tag{5.3}$$

where x_i is the ith component of x. Another commonly used norm, the infinite norm, is defined by

$$\|x\|_\infty = \max_{1 \leq i \leq n} |x_i| \tag{5.4}$$

which is simply the largest component of the vector in magnitude. A third vector norm is the unit norm defined by

$$\|x\|_1 = \sum_{i=1}^{n} |x_i|\,. \tag{5.5}$$

Figure 5.1 shows unit length vectors for all three norms. Although the infinite and unit norms are not as intuitively obvious, they have computational advantages.

Next we define the matrix norm of square matrices A as

$$\|A\| = \max_{x \neq 0} \frac{\|Ax\|}{\|x\|}\,, \tag{5.6}$$

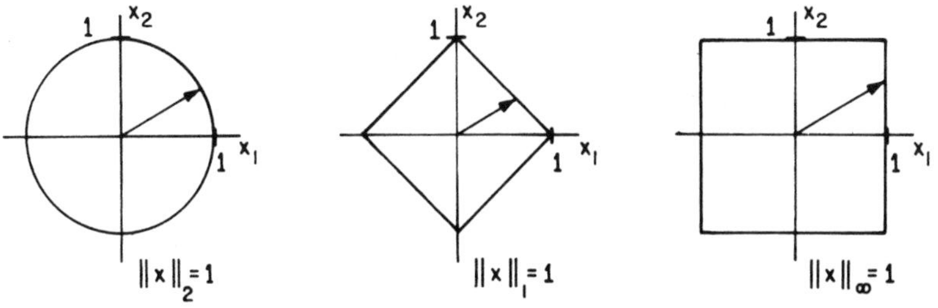

Fig. 5.1. Unit vectors under different norms

or equivalently

$$\|A\| = \max_{\|x\|=1} \|Ax\| . \tag{5.7}$$

Notice that the above equations define a matrix norm in terms of a vector norm. Thus each vector produces a matrix norm that is said to be compatible with the generating vector norm. When the Euclidean norm is used it can be shown that

$$\|A\|_2 = \sqrt{|\lambda_{\max}|} , \tag{5.8}$$

where $\lambda_{\max}$ is the maximum eigenvalue of $A^{\mathrm{H}} A$ in magnitude. For the more familiar case of A being real and symmetric

$$\|A\|_2 = |\bar{\lambda}_{\max}| , \tag{5.9}$$

where $\bar{\lambda}_{\max}$ is the maximum eigenvalue of A. An advantage of the infinite matrix norm is its ease of computation. It can be shown [5.2] that this norm is easily calculated by

$$\|A\|_\infty = \max_{1 \leq i \leq n} \sum_{j=1}^{n} |a_{ij}| . \tag{5.10}$$

Now the condition number of a square matrix, discussed thoroughly by Forsythe and Moler [5.3], can be defined.

$$\operatorname{cond}(A) = \|A\| \cdot \|A^{-1}\| . \tag{5.11}$$

For the Euclidean norm, the condition number can be calculated

$$\operatorname{cond}_2(A) = \sqrt{|\lambda_{\max}/\lambda_{\min}|} , \tag{5.12}$$

where λ_{max} and λ_{min} are the largest and the smallest eigenvalues in magnitude of the Hermitian matrix $A^H A$. For the case in which A is real and symmetric, one can more easily see why a large condition number implies an ill-conditioned problem. For this case

$$\mathrm{cond}_2(A) = \frac{|\bar{\lambda}_{max}|}{|\bar{\lambda}_{min}|}, \tag{5.13}$$

where $\bar{\lambda}_{max}$ and $\bar{\lambda}_{min}$ are maximum and minimum eigenvalues in magnitude of A. One can think of premultiplication by a matrix as a vector function in which components of a vector in the direction of the eigenvectors of the matrix are stretched or shrunk by the respective eigenvalues. Suppose that f in (5.2) has some error δf caused by some error δg in g. Then (5.2) becomes

$$A(f + \delta f) = g + \delta g. \tag{5.14}$$

Now if δf is in the direction of the eigenvector associated with the minimum eigenvalue

$$\bar{\lambda}_{min}\,\delta f = \delta g. \tag{5.15}$$

However, in the inversion process where we calculate f from g, we, in effect, calculate

$$\delta f = \frac{1}{\bar{\lambda}_{min}}\,\delta g. \tag{5.16}$$

Since $\bar{\lambda}_{min}$ is a small number, $|\bar{\lambda}_{max}/\bar{\lambda}_{min}|$ will be a large number. Thus, if the condition number is large, the eigenvalues vary greatly in size, and therefore, the relative error in f may be much larger than the relative error in g. The geometric interpretation for the infinite norm is not as clear.

From the definition of the condition number and elementary properties of matrix and vector norms, one can derive estimates for the error in f due to known errors in g or A. First, consider the case where A is precisely known but g contains some error δg. Then we do not know g but only $g + \delta g$, and thus inversion of (5.2) yields not f but $f + \delta f$ where δf is the resulting error in f. From

$$\delta f = A^{-1}\,\delta g, \tag{5.17}$$

$$\|\delta f\| \leqq \|A^{-1}\| \cdot \|\delta g\| \tag{5.18}$$

follows from the definition of matrix norms. Similarly from (5.2)

$$\|g\| \leqq \|A\| \cdot \|f\| . \tag{5.19}$$

If we multiply (5.18) and (5.19) and divide by $\|f\|$ and $\|g\|$, which we can assume are nonzero, we find

$$\frac{\|\delta f\|}{\|f\|} \leqq \|A\| \cdot \|A^{-1}\| \frac{\|\delta g\|}{\|g\|} \tag{5.20}$$

or

$$\frac{\|\delta f\|}{\|f\|} \leqq \operatorname{cond}(A) \frac{\|\delta g\|}{\|g\|} . \tag{5.21}$$

Thus, relative uncertainty in f is related to relative uncertainty in g by a function of A. Similarly, when g is precisely known but A contains error δA, the resulting error can be estimated by

$$\frac{\|\delta f\|}{\|f+\delta f\|} \leqq \operatorname{cond}(A) \frac{\|\delta A\|}{\|A\|} . \tag{5.22}$$

It should be noted that the above two inequalities are the sharpest inequalities possible, i.e., the equality will occur for some directions of g and δg.

To apply (5.21) and (5.22) one can use $\operatorname{cond}(A)$ based on either norm. Although the use of the Euclidean norm is geometrically clear, it requires far more computation than using the infinite norm. First, one must form the Hermitian matrix $A^{\mathrm{H}} A$ before finding the largest and smallest eigenvalues. This multiplication, in itself, takes on the order of n^3 multiplications. Since only the largest and smallest eigenvalues are needed, one could find these eigenvalues by iteration. Each iteration takes on the order of n^2 multiplications and our experience has been that convergence is slow. In fact, the easiest way to find the minimum eigenvalue appears to be inverting $A^{\mathrm{H}} A$ and taking the reciprocal of the value. Using the infinite norm requires far less work. All that is needed is to find the maximum row sum of A and the maximum row sum of A^{-1}. One can still roughly estimate an error in an Euclidean sense since $\operatorname{cond}_2(A)$ is related to $\operatorname{cond}_\infty(A)$ by

$$\frac{1}{n} \operatorname{cond}_\infty(A) \leqq \operatorname{cond}_2(A) \leqq n \operatorname{cond}_\infty(A) . \tag{5.23}$$

After considering the relative amounts of work involved, one can see that using the infinity norm, which is easy to calculate and gives the error in a maximum component sense, is more practical from a computational viewpoint.

5.1.2. Scattering by Rectangular Conducting Cylinders

We now proceed to illustrate the application of the condition-number concept with several examples. The first problem, scattering by rectangular cylinders, demonstrates several properties of ill-conditioned problems and shows how the condition number can be used to detect situations where the formulation is ill-conditioned and to evaluate various methods for obtaining an acceptable answer in these situations.

The geometry of this problem is shown in Fig. 5.2. A uniform plane wave is incident upon an infinite rectangular cylinder with width W and depth D. Since we are considering the wave to have no variation along the axis of the cylinder, we can consider the problem to be two-dimensional. We wish to calculate both induced currents and scattered fields. For simplicity, only waves normally incident to the front face will be considered.

The incident wave can be decomposed into waves of two separate polarizations also shown in Fig. 5.2. For the Transverse Magnetic (TM) polarization, or "E wave", the H field is perpendicular to the axis of the cylinder and hence the E field is parallel to the axis. Similarly, for the Transverse Electric (TE) polarization, or "H wave", the H field is parallel to the axis.

Results for each polarization are usually calculated separately using separate formulations. The details for the derivations of these formulations have been discussed by many authors (see [5.4] and [5.5]); we will

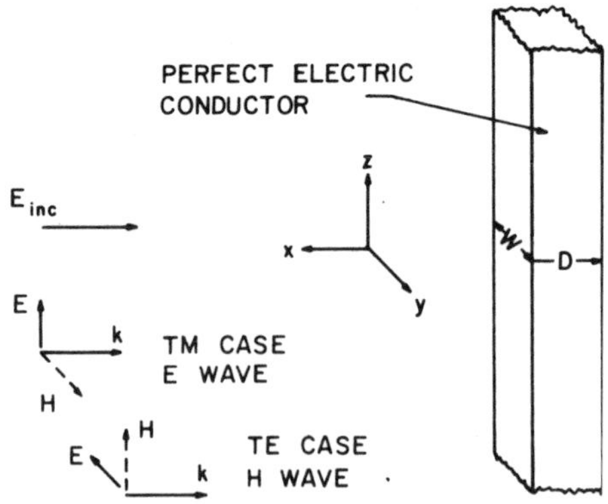

Fig. 5.2. Geometry of rectangular cylinder scattering problem showing the two excitation polarizations

only outline the methods. The usual formulation used for the TM excitation is an E field integral equation which enforces the condition that the total tangential E field be zero along the surface of the cylinder. One can show that the z-directed field due to current I carried by an infinitely long wire, distance r away, is given by

$$E_z = -\frac{k\eta}{4} H_0^{(2)}(kr)\, I\,, \tag{5.24}$$

where k is the free space wavenumber $2\pi/\lambda$, η is the intrinsic impedance of free space, and $H_0^{(2)}$ is the Hankel function of zeroth order, second kind. Applying the condition

$$E_z^{\mathrm{inc}} + E_z^{\mathrm{scat}} = 0 \tag{5.25}$$

along the boundary, where the two terms are the incident and scattered fields respectively, yields

$$\frac{k\eta}{4} \int_C H_0^{(2)}(kr)\, J_z(\varrho)\, dl = E_z^{\mathrm{inc}}(\varrho_s)\,, \tag{5.26}$$

where r is the distance from source to observation point, J_z is the current density along the boundary of the rectangle, and the integral path C is taken over this same rectangular path. After the current has been determined, the scattered field is expressed in the calculation of the bistatic radar cross section, which is given as a function of the angle ϕ from the direction of the incident wave as

$$\sigma(\phi) = \frac{k\eta^2}{4} \left| \int_C e^{jk(x'\cos\phi + y'\sin\phi)}\, J_z(x', y')\, dl \right|^2, \tag{5.27}$$

when the incident E field has unit magnitude.

An H-field integral equation is often used to solve the TE case. For this case excitation currents will be circumferential. Starting with the condition

$$H_z = H_z^{\mathrm{inc}} + H_z^{\mathrm{scat}} \tag{5.28}$$

and recalling $H_z = J_c$ one can derive the integral equation

$$J_c(\varrho_s) + \frac{jk}{4} \int_C (\boldsymbol{a} \cdot \boldsymbol{n})\; \mathrm{H}_1^{(2)}(kr)\, J_c(\varrho)\, dl = \mathrm{H}_z^{\mathrm{inc}}(\varrho_s)\,. \tag{5.29}$$

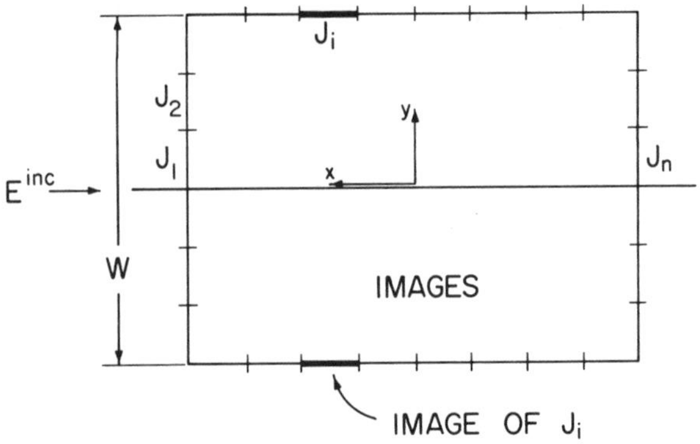

Fig. 5.3. Pulse expansion functions used for representing currents on rectangular cylinder showing symmetry used to reduce matrix order

In these equations H_z is the total z-directed H-field just external to the cylinder, J_c is the circumferential current density, $\mathrm{H}_1^{(2)}$ is the Hankel function of the first order, second kind, $\boldsymbol{n}$ is the outward normal vector along the cylinder, $\boldsymbol{a}$ is a unit vector pointing from the source point to the observation point, and r is the distance between these two points. Analogously to the TM case, the radar cross section can be expressed by

$$\sigma(\phi) = \frac{k}{4} \left| \int_C \boldsymbol{n} \cdot \boldsymbol{R} \mathrm{e}^{jk(x' \cos\phi + y' \sin\phi)} J_c(x', y')\, dl \right|^2 , \tag{5.30}$$

where $\boldsymbol{R}$ is the unit vector in the ϕ direction, and the incident H field has unit magnitude.

For both polarizations the integral equations obtained can be easily discretized by the method of moments. The discretization may be done by dividing the perimeter of the cross section into equal length segments and solving for currents J_i that are uniform over these segments (see Fig. 5.3). Following the technique of point matching we will specify the value of the field at the midpoints of each segment. Since the incident wave has been restricted to be normal to the front face, one can use the resulting symmetry to halve the order of the matrix while keeping the same number of segments (see Fig. 5.3). Applying the symmetry consideration, therefore, is equivalent to using pairs of rectangular currents as the expansion functions. In calculating the matrix elements, one must use a small-argument expansion of the kernel for the self elements; for all the other elements, the approximation that all the current over a segment is concentrated at its midpoint is usually sufficient. An easy way to improve this last approximation is to calculate the fields at the testing point due to currents at the midpoint and the two endpoints of the source segment weighted by the proper Simpson rule coefficients. This improvement was used for the TM case results to be given later.

These formulations described above for the two polarizations yield good results except over a narrow band of certain discrete frequencies where results vary wildly with small changes in frequency. Figures 5.4 and 5.5 show the induced currents on a square cylinder for both polarizations and for both resonant and non-resonant frequencies. For the non-resonant cases the results agree well with those of MEI and VAN BLADEL [5.4]. Notice how the resonance affects the fourth curve. For the TM case, although incorrect currents are obtained, the scattered fields remain correct except within a very much narrower band. For the TE case, both currents and fields are incorrect. Figure 5.6 shows the condition number graphed against the width of the rectangle in wavelengths. The sharp rise in the condition number accurately indicates ill-conditioned regions where incorrect answers occur. The frequencies and nature of the incorrect TM solutions are shown to be related to certain TM waveguide modes at cutoff although certain other modes do not appear. Conversely, one might expect the incorrect solutions of the TE case to be related to TE waveguide modes, but instead it can be seen that the incorrect currents are again related to TM waveguide modes. The explanation for these puzzling results can be found upon re-examining several properties of the formulation that are usually taken for granted with a routine application of the method of moments.

One starts to understand the apparent nonuniqueness of the results at certain frequencies by recalling the theorem that the operator equation

$$L(f)=g \tag{5.31}$$

has a unique solution only if the homogeneous equation

$$L(f)=0 \tag{5.32}$$

has no solution. Clearly, if such a solution f_h to (5.32) does exist, $f+\alpha f_\mathrm{h}$ satisfies (5.31) equally well for any value of α. For the TM case, (5.26) clearly has the form of

$$L(J_z)=E_z \text{ on the boundary}. \tag{5.33}$$

The nonuniqueness in the TM case is evidently caused by the addition of some component of current which causes no tangential E-field on the boundary. Such a current is that supported in the walls of a waveguide

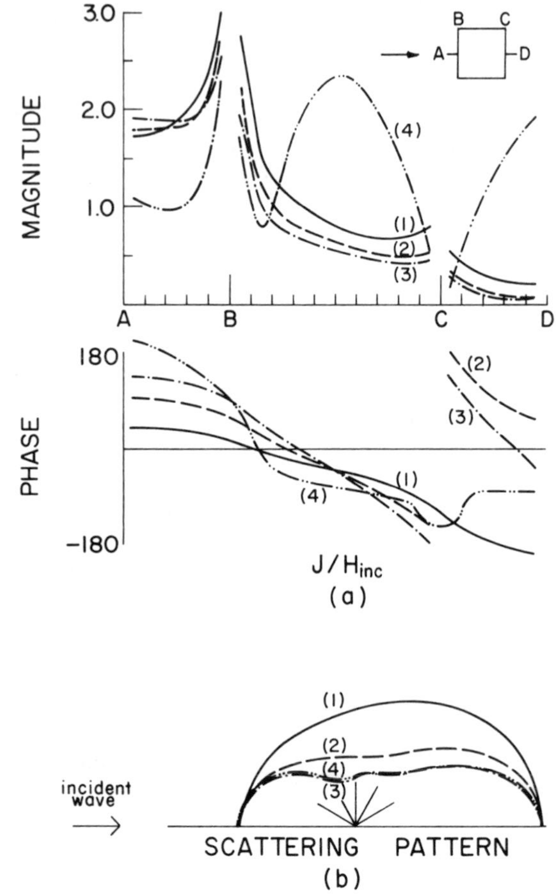

Fig. 5.4. (a) Magnitude and phase of $J/H_{\rm inc}$ for TM excitation of various width square cylinders, $n = 20$. 1) $W = 0.25\,\lambda$, 2) $W = 0.5\,\lambda$, 3) $W = 0.701\,\lambda$ (with internal constraints), 4) $W = 0.707\,\lambda$ (without internal constraints). (b) Normalized E-field scattering patterns produced by corresponding currents

at cutoff when there is no z-variation of a mode. Examination of Fig. 5.4 shows that the incorrect current for a square cylinder has the sinusoidal distribution of a TM_{11} waveguide mode which is resonant at this frequency. Since the resonant waveguide mode current does not radiate, the calculated scattered field is not appreciably affected by the homogeneous solution.

The next question to consider is why only certain resonances appear. The answer is that the choice of expansion functions has restricted all solutions to be symmetric about a line. Hence, only those TM modes

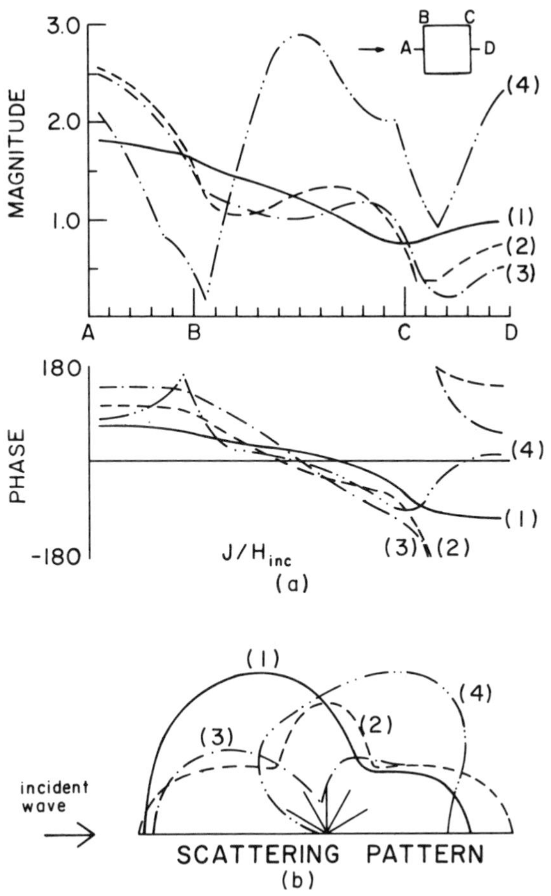

Fig. 5.5. (a) Magnitude and phase of J/H_{inc} for TE excitation of various width square cylinders, $n = 20$. 1) $W = 0.25\,\lambda$, 2) $W = 0.5\,\lambda$, 3) $W = 0.707\,\lambda$ (with internal constraints), 4) $W = 0.707\,\lambda$ (without internal constraints). (b) Normalized E-field scattering patterns produced by corresponding currents

which support currents that have symmetry about this line can exist to cause trouble. These possible modes have an odd number of half-wave variations; hence, only those modes appear which are represented by $TM_{2n+1,m}$ where the first index corresponds to width.

Understanding the nature of the homogeneous solutions for the TE case resonances is slightly harder. The peculiarities present in this case that were not present in the TM case stem from the fact that (5.29) is a H-field integral equation represented by a Fredholm equation of the second kind rather than the first kind. Although both integral Eqs.

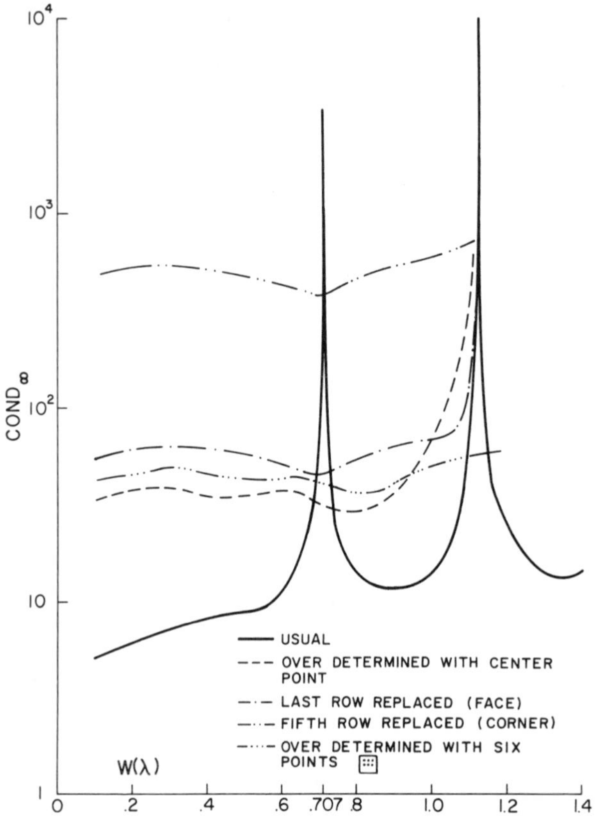

Fig. 5.6. Condition number versus width of square cylinder in wavelengths (TE excitation, $n = 20$)

(5.26) and (5.29) have the form

$$L(J) = F^{\mathrm{inc}}, \tag{5.34}$$

where F^{inc} is the incident field, only for the TM case, does the operator L give the scattered field on the boundary. For the TE case, merely the integral part of L represents the scattered field. Clearly then for the TE case, $L(J)$ can have a homogeneous solution without the scattered field being zero. Next we consider why the homogeneous solution has the functional variation of a TM resonance. Implicit in the formulation of (5.29) is the fact that we are considering fields just exterior to the cylinder which are not continuous with those fields which are just interior. It is also possible to formulate an H-field integral equation for the TM excitation which has identical form to (5.29) with the only difference that one subtracts the integral contribution rather than adding it [5.6]. When this formulation is applied to the interior region, instead of the exterior region, the one minus sign changes and the resulting equation becomes identical in form to (5.29), the TE excitation H-field equation we are using for the exterior. Hence the resonance we experience solving the *exterior* problem for the TE case must have the same variation as the resonance for the *interior* for the TM case. An E-field integral equation does not cause such problems since the E-field is continuous at zero across the boundary.

We have already seen that the condition numbers identify the troublesome frequency regions in agreement with our knowledge of rectangular waveguide modes. In a more arbitrarily shaped waveguide, however, one could not as easily calculate the resonant frequencies and would have to depend on the behavior of the condition number to indicate difficulties. It will next be shown how the condition number can be used in evaluating techniques that may be employed for circumventing these difficulties and deriving accurate results in the presence of ill-conditioning. The three methods to be discussed are filtering, regularization, and extending the boundary conditions.

An intuitive solution to obtaining an acceptable answer to an ill-conditioned problem would appear to be filtering out the unwanted part of the incorrectly calculated answer. For this problem one knows exactly the form of the resonant current at any particular resonant frequency and hence could remove any such component of the answer. One finds, however, that the results from this method are very unsatisfactory. The first reason is that since nothing has been done to the matrix, the formulation is still very unstable to small errors. In particular, the resonant current that has been computed numerically is slightly different from the theoretical resonant current since the theoretical current is a homogeneous solution to the integral equation and hence can exist only at exactly one frequency; in contrast, the numerical resonance occurs over a narrow but finite band. Even if one wished to use the eigenvector associated with the minimum eigenvalue there remains a second, more fundamental problem that part of the resonant solution really should be present. Although forcing the coefficient of the homogeneous solution to zero can produce a unique solution throughout a resonance region the results are clearly incorrect compared with results outside the resonance. Therefore, the major fault of filtering is that no information is provided which can correctly determine the coefficient of the resonant solution.

The second method, regularization, can actually make the matrix more stable. Only a simple scheme of regularization will be discussed here; more general descriptions can be found elsewhere [5.7, 5.8]. By modifying the matrix, one can restrict the possible solutions to be more acceptable; for example, by changing the matrix equation

$$Af = g \tag{5.2}$$

to

$$(A + \delta I)\, f = g\,; \qquad I = \text{identity matrix} \tag{5.35}$$

one adds δ to all eigenvalues and hence almost completely eliminates responses due to eigenvectors associated with eigenvalues that were less than δ in magnitude. Since in this problem only one eigenvalue becomes degenerate (one can think of this eigenvalue causing the condition number to rise at resonances) the regularization parameter δ can be set to discriminate only against the resonant component. As before, however, eliminating the resonant component rather than correctly determining its coefficients, yields a current distribution that is physically unacceptable.

The third method, extending the boundary conditions [5.9], not only makes the matrix stable, but also correctly determines the coefficient of the homogeneous solution. Since we can fill the interior region of the cylinder with a perfect electric conductor without changing currents or scattered fields, restricting the total field to be zero at specified interior points is a valid condition for the fields to satisfy. The matrix equation is modified by using these internal points as testing points in the method of moments. In this way the fields associated with the resonant currents can be constrained since they are zero only on the edge and not in the interior. Using these new testing points in addition to the usual n testing points on the perimeter yields an overdetermined matrix equation which can be solved by applying the usual Moore-Penrose pseudo-inverse for which theory guarantees the minimum-norm, least-squares solution [5.10]. In Fig. 5.6 one can see that the condition number for the extended boundary condition method changes smoothly with frequency and indicates that ill-conditioning is no longer present. Also shown is that increasing the number of internal points from one to six does not seriously degrade any results. Since at resonance one can think of the rows as being nearly linearly dependent in the unmodified method, it is logical to try replacing some row with a new row based on an internal constraint. Although this application of internal constraints is not as versatile as the first, it works surprisingly well even away from resonances. The condition numbers indicate that it is better to replace a row representing the side of a face rather than one representing the corner. One must also be careful that the chosen interior points are not on nodal lines of a modal field or else the problem will be as ill-conditioned as before.

5.1.3. Waveguide Discontinuities

We now wish to look at some further problems which show additional aspects of ill-conditioned problems and demonstrate how the infinite condition number can be of use. In this section we consider the problem of calculating the scattered fields in a bifurcated parallel-plate wave-

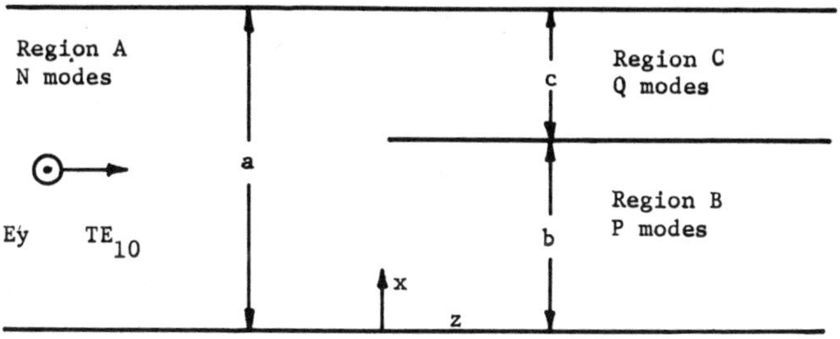

Fig. 5.7. Geometry of bifurcated waveguide problem

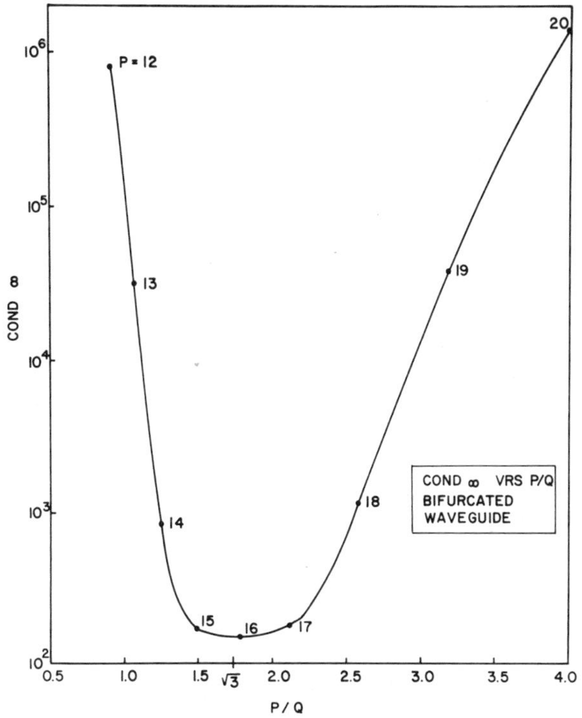

Fig. 5.8. Condition number as a function of the modes representing the fields in Regions B and C. $P+Q=25$

guide [5.11]. This problem was chosen for investigation since an exact solution of the associated integral equation is available. The geometry of the problem is shown in Fig. 5.7. Suppose that a TE_{10} modal field is incident from the left and that we wish to find the scattered fields in Regions A, B, and C. Matching the fields at the boundary, $z=0$, allows one to make an expression involving the modes in region A in terms of the wave numbers of Regions B and C. This gives rise to a doubly infinite set of equations obtained by the mode-matching procedures at the interfaces A–B and A–C. The relative convergence property of this problem is that one must choose the correct ratio of equations from Regions B and C. Suppose there are P equations from B and Q equations from C. Figure 5.8 shows the condition number using the infinite norm as a function of P/Q. The graph shows which choice of P/Q gives the best-conditioned matrix for an order of 25. For this specific problem it has been shown on physical grounds that the ratio of P to Q must be the ratio of the respective region widths. For this example $b/c=\sqrt{3}$; the graphs verify that the best ratio of P to Q is 16/9. However, in other problems showing the relative convergence phenomena, the correct value of P/Q or the correct field values is not known *a priori*. In such a case, a study of the condition number might show the correct choice of P/Q and thus the correct field values.

5.1.4. Wavefront Reconstruction

Another problem analyzed with condition numbers is that of wavefront reconstruction. In this problem, (5.1) becomes the familiar Fourier transform. Upon limiting the range of x to $[-XH, XH]$ and the range

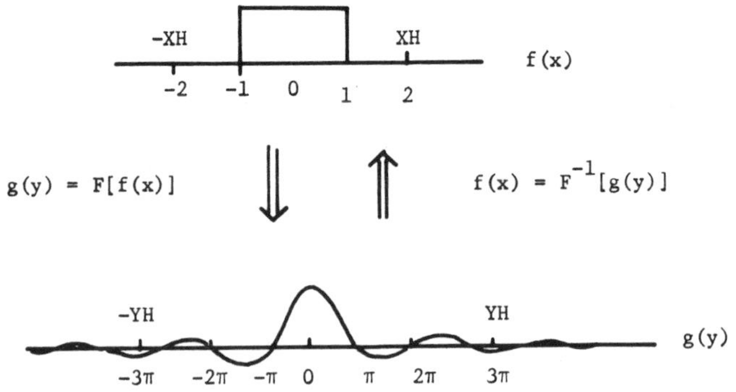

Fig. 5.9. Wavefront reconstruction problem. Deduce $f(x)$ over $[-XH, XH]$ from n samples of $g(y)$ over $[-YH, YH]$

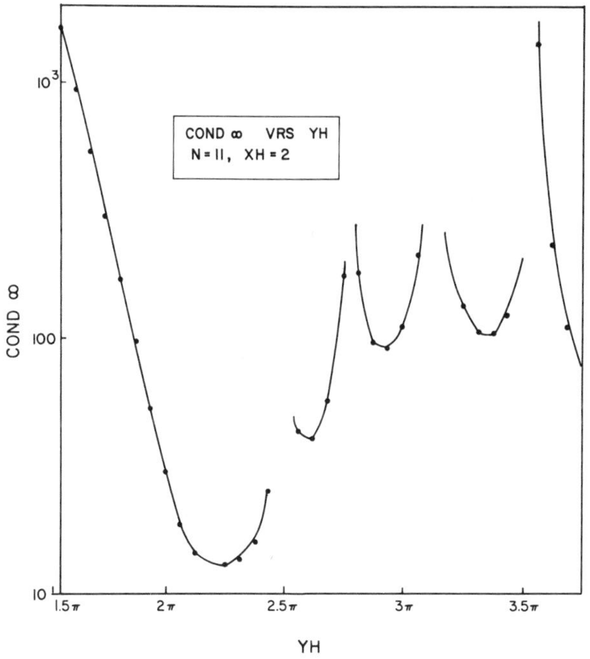

Fig. 5.10. Condition number versus sampling width for wavefront reconstruction problem

of y to $[-YH, YH]$, (5.1) reads

$$\int_{-XH}^{XH} e^{jxy} f(x)\, dx = g(y), \qquad -YH \leqq y \leqq YH. \tag{5.36}$$

Figure 5.9 shows the known functions used for f and g as a test. Variables that can be varied are XH, YH, n, and the choice of expansion functions. Figure 5.10 exhibits the condition number versus the sampling width YH, where XH and n are fixed and where rectangular expansion functions have been used for $f(x)$. One result is that when the condition number is a minimum, the error is a minimum and sample points over $[-YH, YH]$ are taken at the Nyquist rate. This rate is predicted by the sampling theorem where the range of y is unlimited; however, this same sampling rate is optimum when y is limited. Since we know the exact solution to this problem, we can study the relationship of the numerical error to the condition number. Figure 5.11 shows that the actual error and the condition number are well-correlated and that the error level is that expected for the word length of the computer.

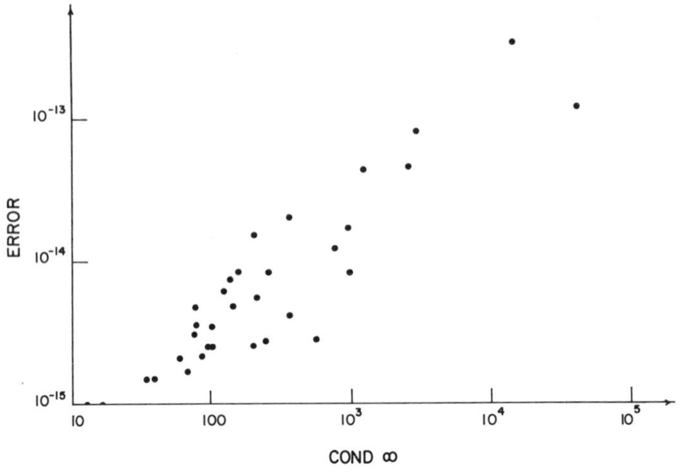

Fig. 5.11. Correlation of error and condition number in wavefront reconstruction problem

5.1.5. Thin-Wire Antennas

Another problem of great interest that has been studied is the determination of current on a thin dipole antenna. Suppose the antenna has a half-length H, radius a, is oriented in the z direction, and is fed in the center with a slice generator. One formulation used is Hallén's integral equation with the approximate kernel

$$\int_{-H}^{H} G(z, z')\, I(z')\, dz' = \frac{-j}{\eta}\left(C_1 \cos kz + \frac{V_t}{2} \sin k|z|\right)$$
$$G(z, z') = \frac{e^{-jkR}}{4\pi R} \qquad R = \sqrt{(z-z')^2 + a^2}\,. \tag{5.37}$$

To solve this equation, one can solve for currents that would be caused by each of the two excitation terms separately and can then determine C_1 so that the total current goes to zero at the end of the antenna. For this equation it is convenient to use rectangular expansion functions and point matching. For Pocklington's equation triangular expansion and testing functions have been used. Figure 5.12 shows that in both formulations the condition increases as n increases. Figure 5.13 exhibits that for the cases studied the conditioning of Hallén's equation varies little with frequency while the Pocklington formulation is strongly affected by frequency and its condition is worse near resonances. However, when the matrix form of Hallén's equation is rearranged so that C_1

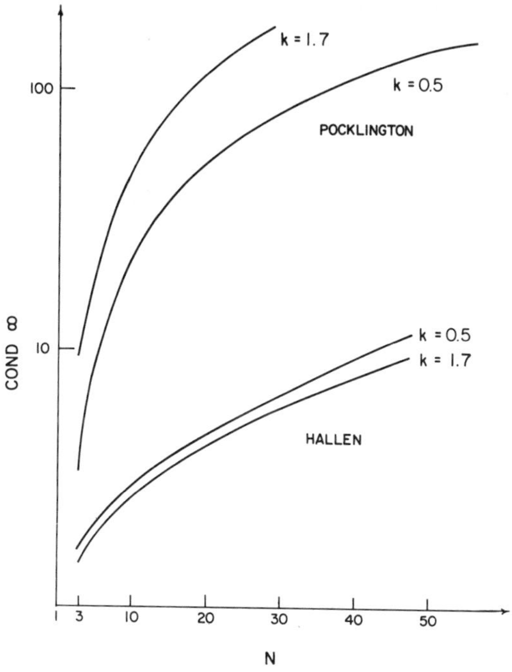

Fig. 5.12. Condition number versus order of matrix for thin-wire dipole problem ($H = 1$)

is one of the original unknowns, the condition number behaves as it did for Pocklington's equation.

Stability calculations have also been made for Pocklington's equation using other expansion and testing functions. Methods that have been compared include triangular expansion and testing functions; piecewise sinusoidal expansion functions with both rectangular testing functions, reaction matching and point matching; and a finite difference method. In all these cases condition numbers were comparable and rose in the same way near resonance. One cannot conclude, from this alone however, that Hallén's equation is necessarily better since there are other factors to consider although for Hallén's equation the instability can be removed from the matrix equation.

5.1.6. Remote Sensing

The example to be considered shows the difficulties of typical remote sensing problems and the necessity of a good choice for expansion functions. The following integral equation arises in finding the dielectric

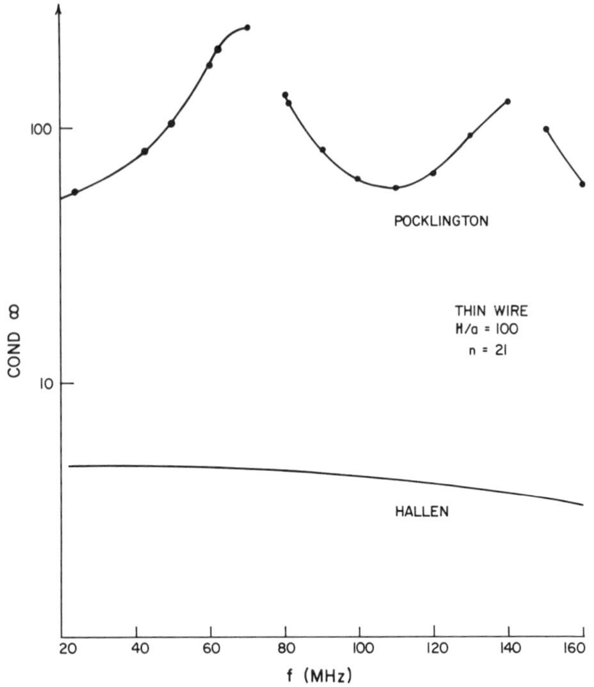

Fig. 5.13. Condition number versus frequency for thin-wire dipole problem ($H = 1$)

profile of a unit thickness slab [5.12],

$$\int_0^1 f(x) \sinh^2 [y(x-1)] \, dx = g(y) . \tag{5.38}$$

Attempts were made to solve this equation first using rectangles and then polynomials as expansion functions. In both cases the condition numbers varied from 100 to 10^{37} depending on n and the range of y. Only in the second formulation, though, were we able to get convergence to the correct answers. However, when 1% random error is added to $g(y)$, which is reasonable for experimental data, the high condition numbers correctly predict that the calculated errors in $f(x)$ will completely obscure the correct answers. Hence, the condition numbers show the feasibility of a particular formulation when experimental error is involved.

It would be very desirable to be able to use the condition numbers as a guide in formulating a problem so that they indicate the best expansion functions to be used. More work remains to be done in this area.

5.2. Convergence

5.2.1. Introduction

Having derived a numerical solution to an integral equation using the moment method, one is often faced with the problem of assessing the accuracy of the solution thus obtained. Typically, this is difficult to accomplish in an absolute sense and one has to appeal to some indirect checks such as comparison with experimental measurements or theoretical solutions for some canonical problems. Although reliable and self-checking tests for the accuracy of numerical solution are difficult to find, it is nevertheless desirable to devise some direct checks that ascertain that the matrix solution indeed satisfies the original integral equation. Further, it is also desirable to establish that the numerical procedure has converged. Both of these topics are examined in this second part of this chapter with the aid of a number of thin wire antenna problems as examples.

To examine how well solutions to antenna problems actually satisfy the boundary conditions and how well various parameters converge when increasing the order of the matrix, we will solve Pocklington's integral equation using piecewise sinusoidal expansion functions and pulse or rectangular testing functions. Pocklington's integral equation is

$$\left(\frac{d^2}{dz^2}+k^2\right)\int_{-H}^{H} I(z')\,G(z,z')\,dz' = -j\omega\varepsilon E_z^{\text{inc}}, \tag{5.39}$$

where one is solving for the current distribution $I(z')$ over the surface of an antenna directed in the z direction with length $2H$. Strictly speaking, one should use the exact kernel

$$G(z,z') = \frac{1}{8\pi^2}\int_0^{2\pi}\frac{e^{-jkR}}{R}\,d\phi' \tag{5.40}$$

with

$$R = \sqrt{(z-z')^2 + 2a^2(1-\cos\phi')}\,,$$

where a is the radius of the antenna, but in most applications the approximate kernel

$$G(z,z') = \frac{1}{4\pi}\,\frac{e^{-jkR}}{R}, \qquad R = \sqrt{(z-z')^2 + a^2} \tag{5.41}$$

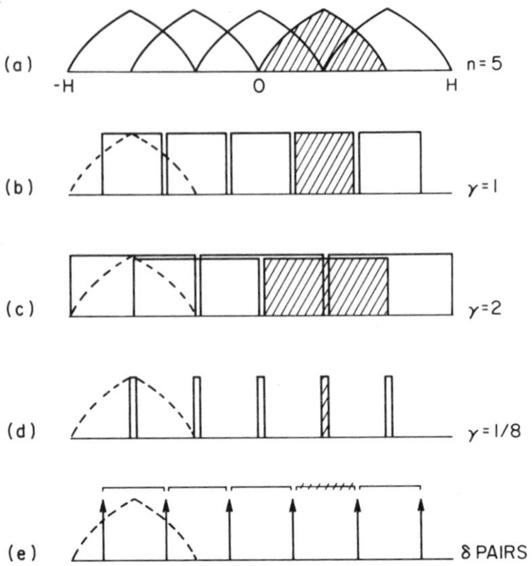

Fig. 5.14a–e. Expansion and testing functions used, shown for $n=5$

is used. The easiest way to view piecewise sinusoids as expansion functions is to consider the current to be the sum of sinusoidal "triangles" shown in Fig. 5.14a multiplied by the coefficients of the current vector one is solving for. The form for the ith expansion function centered about z_1 is simply

$$I_i = \frac{\sin[k(\Delta z - |z - z_i|)]}{\sin k\Delta z}, \quad |z - z_i| \leqq \Delta z \qquad (5.42)$$
$$= 0 \quad , \quad \text{otherwise},$$

where the width of the function is $2\Delta z$. One of the advantages of these expansion functions is that the electric field can be easily calculated in terms of distances from discontinuities in the slope of the current. The parallel component of the field produced by one sinusoidal triangle is

$$E_z = \frac{30j}{\sin k\Delta z}\left[\frac{e^{-jkR_1}}{R_1} + \frac{e^{-jkR_2}}{R_2} - 2\cos(k\Delta z)\frac{e^{-jkR}}{R}\right], \qquad (5.43)$$

where R_1 and R_2 are the distances from the observation point to the endpoints of the interval, and R is the distance to the midpoint of the interval [5.13].

5.2.2. Testing Functions

For the testing functions one of the simplest and most convenient functions is a sequence of pulses centered over the corresponding expansion functions. Let us specify the width of these pulses to be $\gamma \Delta z$, where γ is a relative width factor that can be varied. When γ is unity the testing functions are Δz wide and appear, as shown in Fig. 5.14b. When γ is increased to 2 (Fig. 5.14c), the testing functions cover the width of the corresponding expansion functions and overlap considerably; however, when γ is decreased to 0.125 (Fig. 5.14c), the rectangles are entirely disjoint. By decreasing γ still farther one can generate the same results as obtained by point matching.

As a result of using uniform spacing on a linear wire, the matrix obtained is symmetric and has the property that all elements a_{ij} for which $i-j$ is a constant are the same. A matrix with this last property is said to have a Toeplitz form. The advantage of our matrix being a symmetric Toeplitz form is not only that merely n unique elements need to be calculated but also that one can use a special algorithm for the inversion which reduces the number of operations to the order of n^2 rather than n^3 [5.14]. In addition, if one is only solving for one excitation and does not need the explicit inverse, the storage space can be reduced from n^2 to $5n$. Hence, because of the symmetric Toeplitz form, larger systems can be easily solved.

Before calculating results for a transmitting antenna one must adequately model the feed region. The slice generator or δ-source model is often used not only because it allows one to rigorously discuss the applied voltage of an antenna which cannot be defined for a distributed source, but also because it is easy to apply. Since the elements of the vector g are the integrated fields over the widths of the corresponding testing functions, the center element of g is simply the one volt excitation and the other elements are zero. However, there are many fields which will have the same vector representation. For example, we can also consider the incident field to be a uniform pulse Δz wide with intensity $1/\Delta z$ and obtain the same results. Understanding then that the integrated field at the feed segment is one, and that this stipulation does not exactly specify what feed model is being used, we can go on to look at results.

We observe the effect of changing γ, the width of the testing function. The first effect is that, as γ decreases, the impedance appears to be approximately γ times what it should be. This effect is inconsistent with the δ source model since one should have obtained complete contribution of the δ source even with the smaller width. It is consistent with the Δz width pulse feed since when γ is smaller one is not sampling the entire pulse and hence must scale results by γ, the amount of overlap between

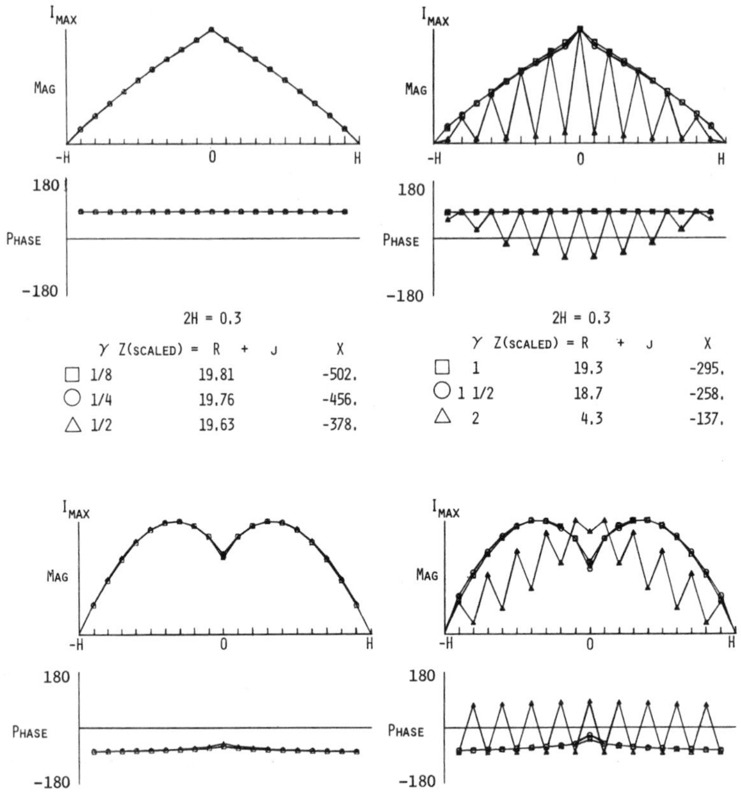

2H = 0.3

γ	Z(SCALED) = R	+ J X
□ 1/8	19.81	-502.
○ 1/4	19.76	-456.
△ 1/2	19.63	-378.

2H = 0.3

γ	Z(SCALED) = R	+ J X
□ 1	19.3	-295.
○ 1 1/2	18.7	-258.
△ 2	4.3	-137.

2H = 0.7

γ	Z(SCALED) = R	+ J X
□ 1/8	333.	606.
○ 1/4	343.	553.
△ 1/2	369.	461.

2H = 0.7

γ	Z(SCALED) = R	+ J X
□ 1	426.	340.
○ 1 1/2	512.	254.
△ 2	11.8	-104.

Fig. 5.15. Current distribution and impedance for different width testing functions ($H/a = 100$, $n = 19$)

the Δz pulse and the testing function. Figure 5.15 shows the current distribution and the resulting impedances scaled by γ for two different length antennas. In this and the following figures, unless otherwise noted, the ratio of the half-length to radius H/a is 100. It can be seen that for both lengths the distributions are almost exactly the same shape for γ's which are one or less. Also, the changes in impedance are not greater than one would expect to find in comparing point-matching with rectangular testing functions. However, for the case $\gamma = 2$ the wildly oscillatory currents and extreme values for impedance clearly indicate that the results are erroneous.

There is no clear *a priori* reason why a rectangular testing function $2\Delta z$ wide should give such poorer results than for one Δz wide. One might wonder whether the matrix has somehow become singular, but a calculation of the matrix condition number indicates that the matrix is actually slightly less ill-conditioned than for $\gamma=1$. Also one might consider whether if one assumes the incident field is a pulse rather than a delta that its vector representation should be $[0, 0, \ldots, 0.5,\ 1,\ 0.5,\ 0, \ldots, 0]$ rather than having just a single 1. This has been tried and results are still oscillatory. In fact, the problem here cannot be the feed modeling at all because oscillatory results remain even when one uses plane wave excitation. It can be noted that the same type of poor results are obtained when one uses pairs of deltas for testing, which is a two-point sampled version of the case $\gamma=2$, as shown in Fig. 5.14e [5.15].

To test the solution's validity and accuracy it will be useful to calculate the near fields actually produced by the calculated currents. Although the true solution should have zero tangential field on the surface, one calculated using the constraints produced by the n testing functions will not. The deviation from zero will indicate how well the calculated currents actually satisfy the boundary conditions. In addition, the field calculations near the feed will show the feed distribution one has been implicity using.

Figure 5.16 shows the near fields for various widths of the testing rectangles. Notice that in addition to the large field at the feed which we expect there is also a large field at the end of the antenna caused by the discontinuity of the slope. Also rather than speculating on whether one should use a delta or pulse for the feed model one can see the actual field implied by a single 1 in the excitation vector. For $\gamma=1$ the shape of the field distribution is the same as that produced by reaction matching [5.16]. The distribution and the large magnitude of the field where $\gamma=2$ indicate that although the integral of the field is zero over the proper intervals, the boundary condition that the field be zero, clearly has not been satisfied.

Figure 5.17 shows the distribution of the field away from the feed in more detail. One can see that the shapes for $\gamma=1/8$ and $\gamma=1$ are almost exactly the same with the only difference being that adjusting the width over which the field must be zero scales the curve and shifts the average value of the field.

In conclusion, we have seen that the choice of testing functions cannot be arbitrary, and that certain choices yield erroneous results. To verify the validity of a moment method solution it is useful to calculate the near fields which show if the boundary conditions have been satisfied.

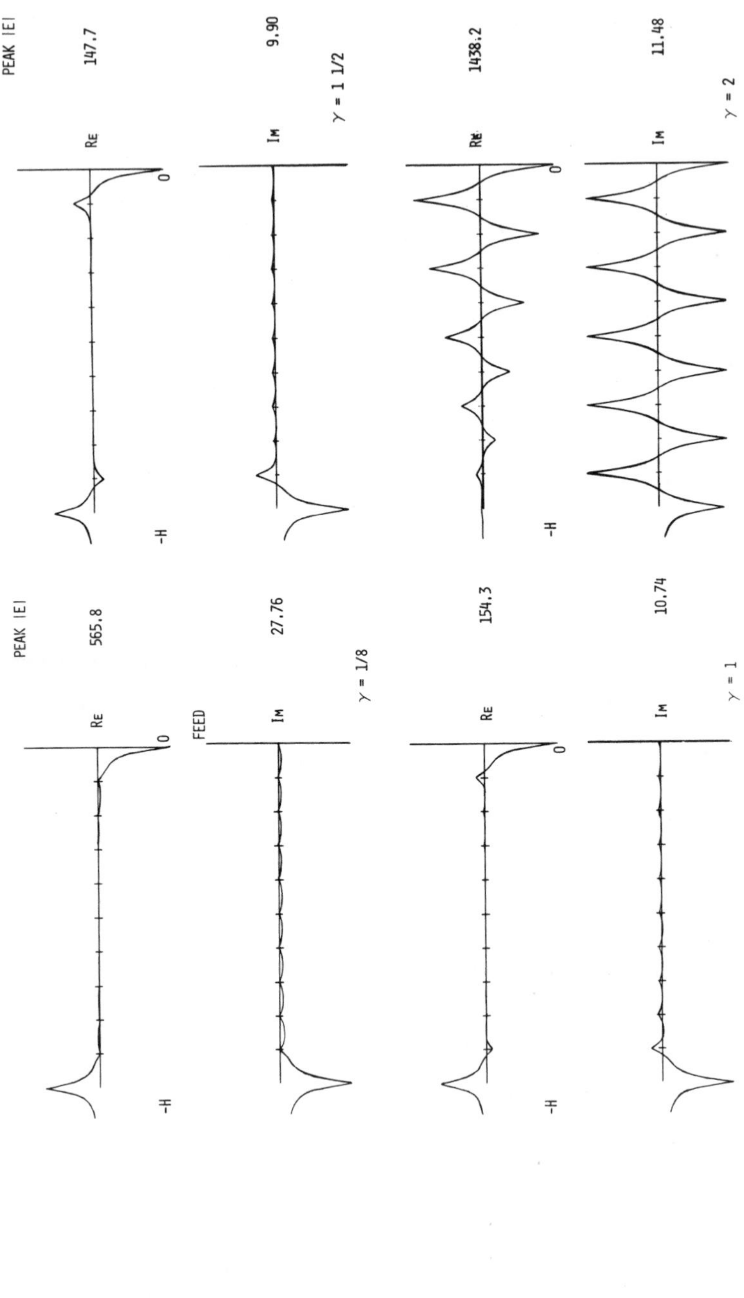

Fig. 5.16. Calculated tangential E-field on surface of antenna and on one segment past the end for different width testing functions. Only left half of antenna shown ($2H = 0.3\,\lambda$, $n = 19$)

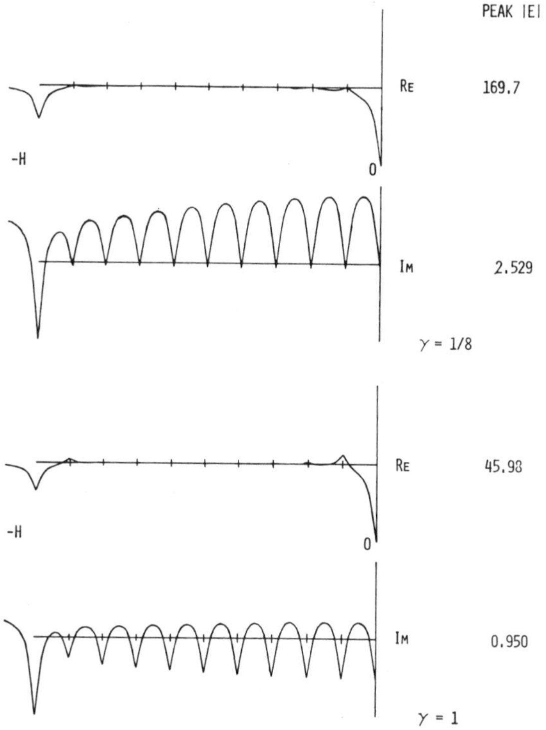

Fig. 5.17. Calculated tangential E field on surface of antenna and one segment off the end ($2H = 1\,\lambda$, $n = 19$)

5.2.3. Convergence of Impedance and Admittance

Once one has obtained reasonable numerical answers the natural inclination is to increase the number of unknowns to find a better answer. Recalling some mathematical theorems we would hope results are converging to an exact limit that would be obtained by using an infinite number of unknowns. However, because the conditions for the theorems do not strictly apply and because of the approximations that are often used, the convergence of numerical results must be seriously questioned.

In discussing convergence one must be aware of several points. Although convergence, in the sense of the limit of a sequence, has a strict mathematical definition, the term convergence is also used more loosely. One has often heard some discussion like Procedure B converges faster than Procedure A since B yields the same results using twenty unknowns that requires forty unknowns for A, although neither procedure might actually converge. We will try to be careful to avoid this

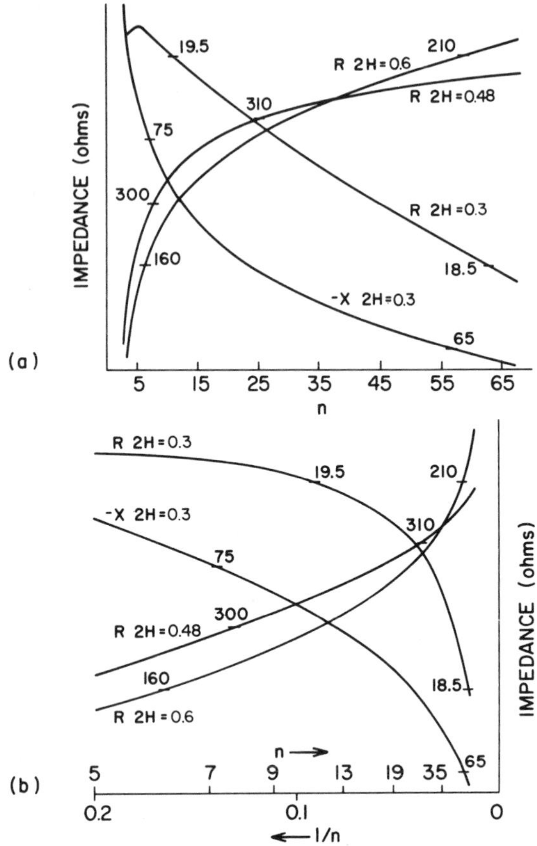

Fig. 5.18a and b. Convergence of antenna impedances (resistance R and reactance X) plotted against n and $1/n$

usage. A second difficulty involves the use of graphical data. Slowly varying functions such as the logarithm function can be scaled to make them appear as if they are converging. Figure 5.18a shows real and imaginary parts of the antenna impedance plotted against n for several length antennas. A better method of graphing for evaluating convergence is to plot the data against $1/n$ so that if the data are converging one can actually graphically estimate the limit. When the data of Fig. 5.18a are replotted this way in Part b, it becomes clear that the impedances are not converging. A third consideration on convergence is that one must remember the physical significance of the quantities being studied. Clearly it is academic to worry about the convergence or divergence of the seventh significant figure if the quantity itself cannot be measured more actually than one percent.

Unfortunately the divergence of the impedance shown in Fig. 5.18b is typical and although here piecewise sinusoids were used for expansion functions and rectangles for testing functions, this same tendency has been noted for other choices. Since results from the method of moments do converge for certain other problems, one may wonder whether the thin-wire approximation (5.41) or the simple modeling of the feed causes this problem to not converge. In an effort to improve convergence, the first method will be to use the exact kernel (5.40). Numerically this can be done by dividing the cylinder into m parallel strips, as given by IMBRIALE (see this volume, Chapter 2) summing the fields from each, and increasing m until the integral converges. A second method will involve keeping the gap width constant. Figure 5.19 shows how by tripling the number of unknowns for a given n and using 1/3 V on each of three segments and continuing similarly for higher odd integers, one can model a uniform E field over a gap with the same physical size. Other methods for improving the convergence use information on the resulting electric field rather than changing the assumptions used in calculating the matrix. A technique developed by MILLER (see this volume, Chapter 4) basically consists of using $-\int E \cdot dl$ over the middle half of the antenna as the excitation voltage rather than the usual assumed 1 V. This technique attempts to compensate for the fact that the calculated field is not confined entirely to the source region. This method is not rigorously valid if the source field distribution is wide; however, the stationary form [5.17]

$$Z = \frac{-\int E \cdot J}{I_{\text{feed}}^2} \, dl \tag{5.44}$$

is valid and has also been computed.

Figure 5.20 shows the radiation resistance for a thin antenna off resonance for the various methods of "improving" convergence. For $\gamma = 1$ Miller's method gives exactly the same results as the usual method since for this choice of testing function the integral of the electric field is automatically forced to be 1. The results from the stationary form for both values of γ shown are above the corresponding usual curves. Recalling the loose definition of convergence, one can note that for a given value of n, the stationary form gives the resistance that requires a much greater value for n using the usual method; however, the curve's shape indicates it is diverging similarly to the usual method. When the feed gap is kept 1/12 of length, it appears that results may actually converge although the usual method and the variational forms give different limits, both of which disagree with other published results.

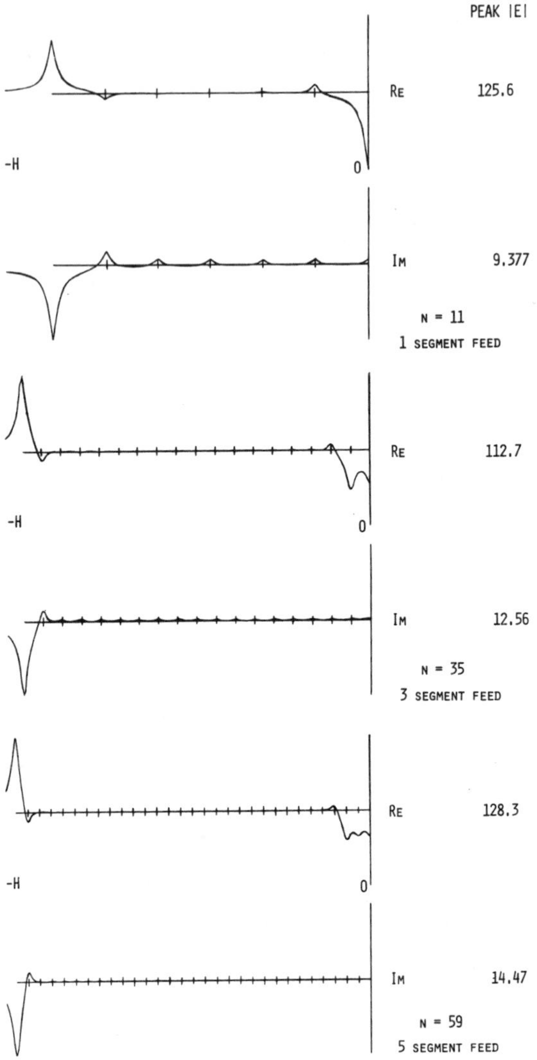

Fig. 5.19. Tangential E-field shows that by increasing the number of unknowns and feed segments one can approach a pulse feed ($2H = 0.3\,\lambda$, $\gamma = 1$)

Conductance values appear to converge better than resistance values. Figure 5.21 shows conductance calculated for the same cases as for the resistance. Where $\gamma = 1$ one can note how the conductance calculated with the exact kernel appears to converge while the thin-wire approximation diverges. Also note how similar the results are

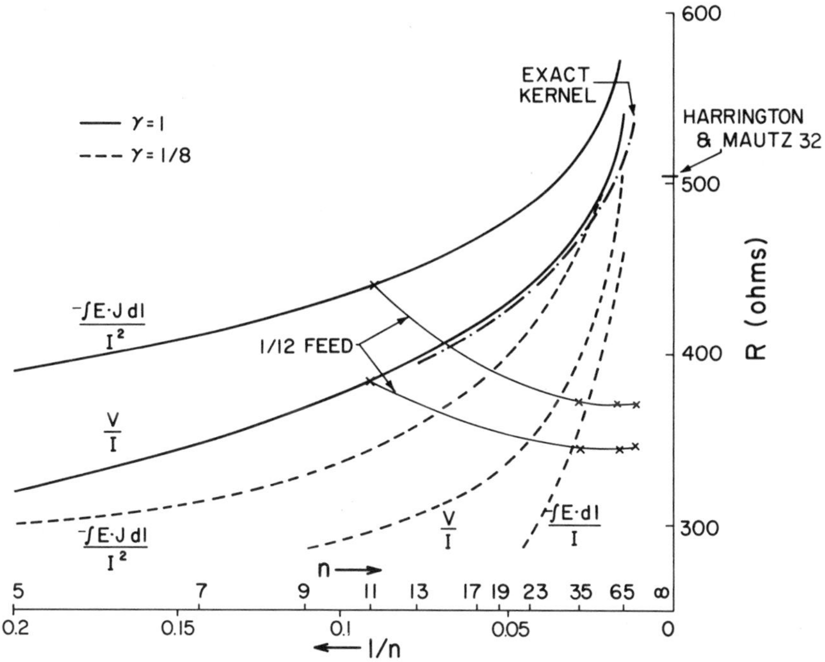

Fig. 5.20. Radiation resistance calculated various ways plotted against $1/n$ ($2H = 0.7\,\lambda$, $H/a = 100$). All curves based on the thin-wire approximation except for curve indicated exact

to the case when the feed gap is kept constant and how results from the usual and variational forms approach each other.

It is somewhat puzzling at first that resistance should diverge while conductance converges. Separating the effect of the thin-wire approximation and the feed model, it appears that the conductance is insensitive to changes in the feed model. This has been tested by fixing the number of unknowns and changing the number of segments used to model the feed thus changing the width of the feed. These changes represent actual physical changes in an antenna and hence one does not expect to get the same answers. The resistance changes considerably but the conductance is virtually constant. This paradox can be resolved in terms of the power radiated in the far field. It is easily shown that conductance should converge as n increases since the power in the far field is not greatly affected by the gap width of a constant voltage source. Modeling the effect of the changing source width as a susceptance in shunt with the true admittance, one can see the gap does not change the real part of the current and hence the power. The gap susceptance enters directly into

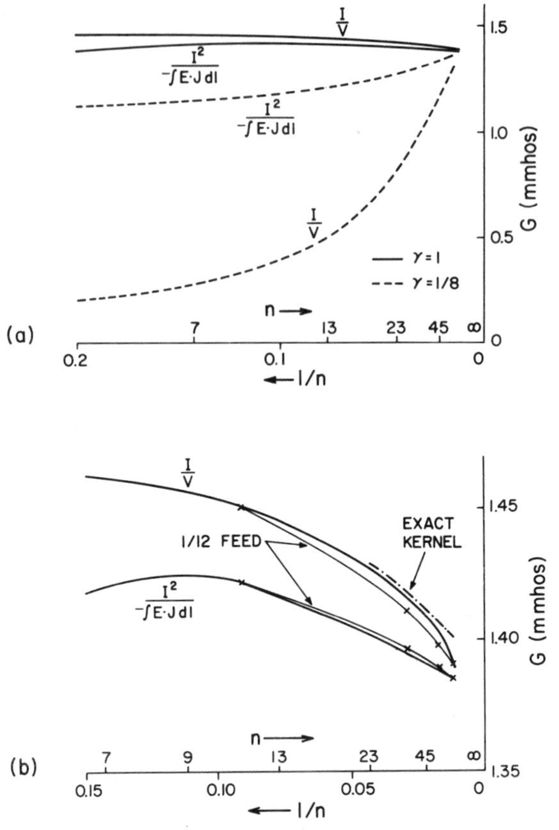

Fig. 5.21a and b. Conductance calculated various ways plotted against $1/n$ ($2H = 0.7\,\lambda$, $H/a = 100$). (a) shows results for different values of γ, (b) shows results for $\gamma = 1$ in more detail

the calculation of the resistance, however, and hence the power radiated from the antenna with a constant current source is not stationary with respect to changing gap width and consequently resistance does not converge with n.

Incidentally, it is now clear why resistances should converge better at resonances. At these frequencies there is no reactive part, thus $R = 1/G$ and if G converges R must also converge.

5.2.4. Convergence of Non-Local Parameters

The computed value of impedance and admittance of an antenna is clearly a local function of the currents and fields at the feeds region. We

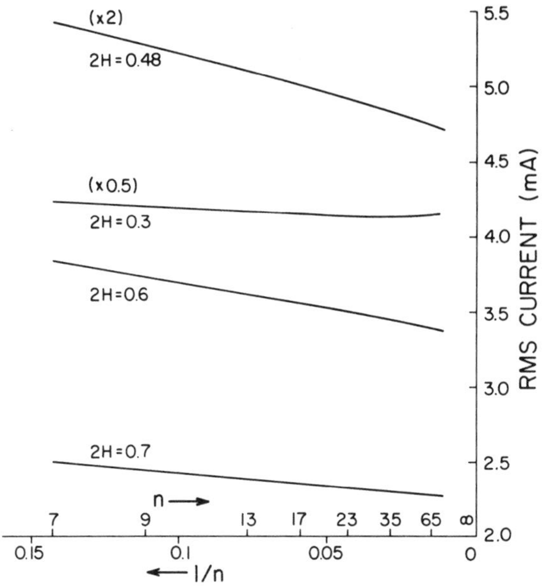

Fig. 5.22. RMS value of current against $1/n$ for various length antennas. Note for $2H = 0.48\,\lambda$ and $0.3\,\lambda$ the true value is the indicated factor times the plotted value

next consider the convergence of a non-local parameter, namely the root-mean-square value of the currents and the fields. The value of current is calculated by

$$\mathrm{RMS}(I) = \sqrt{\frac{\sum_{i=1}^{n} |I_i|^2}{n}}, \tag{5.45}$$

where I_i's are the elements of the current vector. The RMS value for the E field is calculated similarly but using 10 values for the field per segment. Figure 5.22 shows that the RMS value of the current apparently does converge.

Suppose we now consider the RMS value of the E-field and study its convergence properties. When the field is investigated over the entire antenna except for the one segment over which the incident field was presumed to exist, the RMS value of E-field actually grows instead of going to zero. The effect is due to the strong fields at the end of the antenna, the fields that average to zero over segments adjacent to the feed segment, and the fact that, as the feed segment decreases in size, the field magnitude must increase to keep the same voltage. If instead,

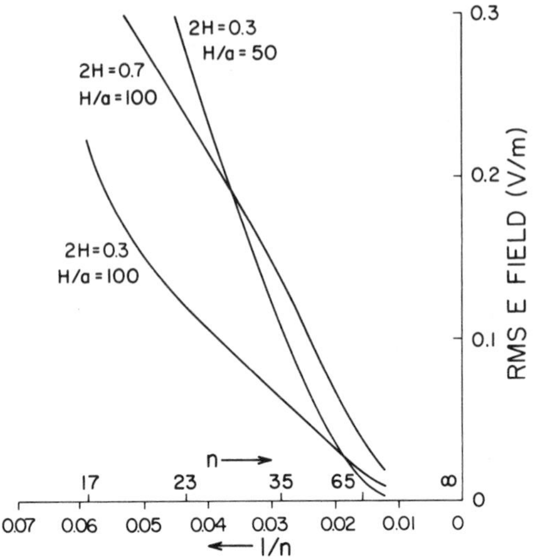

Fig. 5.23. RMS value of E-field over one segment halfway between the endpoint and feed of the antenna, plotted against $1/n$

one considers the RMS value of the E-field over one segment approximately halfway between the feed and the end, one obtains the results shown in Fig. 5.23 which shows that field converges to zero quite quickly.

5.3. Conclusions

The discussions presented in this chapter bring out the fact that it is in general necessary to appeal to certain tests of stability and convergence in order to ascertain that a reliable solution to the original integral equation describing the boundary value problem has indeed been determined via the numerical technique. The "condition number" has been found to be a useful guide for detecting the regimes where the solution may be unstable and where caution must be exercised, or special techniques must be followed to derive stable solutions.

It has been demonstrated that the choice of certain combinations of basis and testing functions in the moment method may result in an erroneous solution to the original problem even when the resulting matrix equation is stable and the equation has been solved with adequate numerical accuracy. It is pointed out that the clue to detect such anomalies lies in the study of the near-field behavior.

Various convergence tests for numerical solution were studied. These investigations shed considerable light onto the convergence of behavior quantities like the input impedance or admittance computed from different formulas. It was found that certain parameters converge better than others. In particular, the conductance appears to be less sensitive to uncertainty in the feed modeling and hence converges better than the resistance. Parameters of the antenna that do not depend on the value of the current at only one point but instead weight all the values converge well. Hence one expects to obtain better converging answers for far fields scattered or radiated by the antenna than for the impedances. It has been shown that the behavior of the E field on the surface of the antenna derived from the computed currents provides an understanding of the accuracy and reliability of the results.

References

5.1. C. KLEIN, R. MITTRA: IEEE Trans. Antennas Propagation AP-**21**, 902 (1973).
5.2. V. N. FADDEEVA: *Computational Methods of Linear Algebra* (translated by C. D. BENSTER from the Russian book of 1950) (Dover Publications, New York, 1959), pp. 54–62.
5.3. G. E. FORSYTHE, C. B. MOLER: *Computer Solution of Linear Algebraic Systems* (Prentice-Hall, Englewood Cliffs, N.J., 1967).
5.4. R. F. HARRINGTON: *Field Computation by Moment Methods* (The Macmillan Company, New York, 1968), pp. 41–61.
5.5. K. MEI, J. VAN BLADEL: IEEE Trans. Antennas Propagation AP-**11**, 185 (1963).
5.6. J. C. BOLOMEY, W. TABBARA: IEEE Trans. Antennas Propagation AP-**21**, 356 (1973).
5.7. G. A. DESCHAMPS, H. S. CABAYAN: IEEE Trans. Antennas Propagation AP-**20**, 268 (1972).
5.8. A. N. TIHONOV: Sov. Math. **4**, 1035 (1964).
5.9. P. C. WATERMAN: Proc. IEEE **53**, 805 (1965).
5.10. C. R. RAO, S. K. MITRA: *Generalized Inverse of Matrices and its Applications* (Wiley and Sons, New York, 1971).
5.11. R. MITTRA, S. W. LEE: *Analytical Techniques in the Theory of Guided Waves* (The Macmillan Company, New York, 1971), pp. 30–37.
5.12. R. MITTRA, D. H. SCHAUBERT, M. MOSTAFAVI: "Some methods for determining the profile functions of inhomogeneous media"; NASA Tech. Memorandum X-62, 150, Mathematics of Profile Inversion (August 1972), pp. 8.2–8.12.
5.13. E. C. JORDAN, K. G. BALMAIN: *Electromagnetic Waves and Radiating Systems* (Prentice-Hall, Englewood Cliffs, N.J., 1968), pp. 333–338.
5.14. D. H. PREIS: IEEE Trans. Antennas Propagation AP-**20**, 204 (1972).
5.15. R. C. MENENDEZ: Private communications.
5.16. G. A. THIELE: In *Computer Techniques for Electromagnetics*, ed. by R. MITTRA (Pergamon Press, New York, 1973), p. 29.
5.17. R. F. HARRINGTON: *Time-Harmonic Electromagnetic Fields* (McGraw-Hill, New York, 1961), pp. 348–355.

6. The Geometrical Theory of Diffraction and Its Application

R. G. KOUYOUMJIAN

With 15 Figures

The rigorous treatment of the diffraction from radiating systems (scatterers and antennas) using the eigenfunction method or the method of Rayleigh, which is based upon the expansion of the field in inverse powers of wavelength, is limited to those objects whose surfaces coincide with the surfaces of orthogonal curvilinear coordinates. Moreover, the solutions obtained are poorly convergent for objects more than a wavelength or so in extent.

Recently there has been considerable interest in the integral-equation formulation of the radiation problem, and its solution by the moment method. Arbitrary shapes can be handled by this method, but in general numerical results also are restricted to objects not large in terms of a wavelength, because of the limitations of present-day computers.

When a radiating object is large in terms of a wavelength the scattering and diffraction is found to be essentially a local phenomenon identifiable with specific parts of the object, e.g., points of specular reflection, shadow boundaries, and edges. The high-frequency approach to be discussed in this chapter employs rays in a systematic way to describe this phenomenon. It was originally developed by KELLER and his associates at the Courant Institute of Mathematical Sciences. This method referred to as the geometrical theory of diffraction is approximate in nature, but in many examples it appears to yield the leading terms in the asymptotic high-frequency solution. Moreover, in many cases it works surprisingly well on radiating objects as small as a wavelength or so in extent. Thus if a solution is desired over a wide spectral range, this high-frequency method nicely complements the low-frequency methods described in the second and third chapters. Finally it will be seen to be sufficiently flexible so that it can be combined with the moment method thereby extending the class of solutions for either method used separately.

The treatment of high-frequency diffraction to follow is restricted to perfectly conducting objects located in isotropic, homogeneous media. The method presented, however, can be extended to penetrable objects and to inhomogeneous and anisotropic media.

6.1. Asymptotic Solution of Maxwell's Equations

6.1.1. Geometrical Optics

Let us begin by examining an asymptotic high-frequency solution to Maxwell's equations in a source-free region occupied by an isotropic, homogeneous medium. Our approach follows that introduced by LUNEBERG [6.1] and KLINE [6.2, 3]. From Maxwell's equations the electric field is found to satisfy

$$\nabla^2 \boldsymbol{E} + k^2 \boldsymbol{E} = 0 \tag{6.1}$$

subject to the condition that

$$\nabla \cdot \boldsymbol{E} = 0 . \tag{6.2}$$

The phase constant $k = \omega\sqrt{\varepsilon\mu}$, where ω is the angular frequency, ε is the permittivity of the medium, and μ is the permeability; a time dependence of $\exp(j\omega t)$ is assumed.

The Luneberg-Kline expansion of the electric field for large ω is

$$\boldsymbol{E}(\boldsymbol{R}, \omega) = \mathrm{e}^{-jk_0\psi(\boldsymbol{R})} \sum_{m=0}^{\infty} \frac{\boldsymbol{E}_m(\boldsymbol{R})}{(j\omega)^m} \tag{6.3}$$

in which $\boldsymbol{R}$ is the position vector, and k_0 is the phase constant of empty space. Substituting (6.3) into (6.1) and (6.2) and equating like powers of ω, one obtains the eikonal equation

$$|\nabla\psi|^2 = n^2 \tag{6.4}$$

together with the first-order transport and conditional equations

$$\frac{\partial \boldsymbol{E}_0}{\partial s} + \frac{1}{2}\left(\frac{\nabla^2\psi}{n}\right)\boldsymbol{E}_0 = 0 , \tag{6.5}$$

$$\hat{s} \cdot \boldsymbol{E}_0 = 0 , \tag{6.6}$$

plus higher-order transport and conditional equations which do not concern us here. In the preceding equations $\hat{s} = \nabla\psi/n$ is a unit vector in the direction of the ray path, and s is the distance along the ray path.

We are interested here in the solution at the high-frequency limit, so the asymptotic approximation for $\boldsymbol{E}$ reduces to

$$\boldsymbol{E}(s) \sim \mathrm{e}^{-jk_0\psi(s)} \boldsymbol{E}_0(s) . \tag{6.7}$$

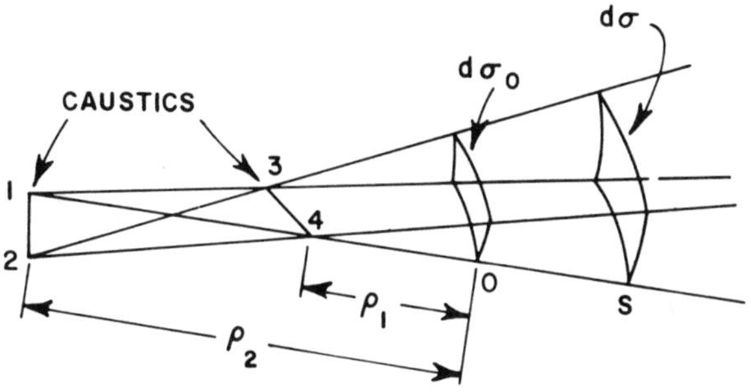

Fig. 6.1. Astigmatic tube of rays

Equation (6.5) is readily integrated and after some manipulation one obtains [6.4, 5].

$$\boldsymbol{E}(s) \sim \boldsymbol{E}_0(0)\,\mathrm{e}^{-jk_0\psi(0)}\sqrt{\frac{\varrho_1\varrho_2}{(\varrho_1+s)(\varrho_2+s)}}\,\mathrm{e}^{-jks} \tag{6.8}$$

in which $s=0$ is taken as a reference point on the ray path, and ϱ_1, ϱ_2 are the principal radii of curvature of the wavefront at $s=0$. In Fig. 6.1 ϱ_1 and ϱ_2 are shown in relationship to the rays and wavefronts.

Equation (6.8) is commonly referred to as the geometrical-optics field, because it could have been determined in part from classical geometrical optics. Specifically the quantity under the square root, the divergence factor, follows from conservation of power in a tube of rays; in addition, we note that the eikonal equation could have been deduced from Fermat's principle, a fundamental postulate of classical geometrical optics. As is well known, classical geometrical optics ignores the polarization and wave nature of the electromagnetic field; however, the leading term in the Luneberg-Kline asymptotic expansion is seen to contain this missing information.

It is apparent that when $s=-\varrho_1$ or $-\varrho_2$, $\boldsymbol{E}(s)$ given by (6.8) becomes infinite, so that this asymptotic approximation is no longer valid. The intersection of the rays at Lines 1–2 and 3–4 of the astigmatic tube of rays is called a caustic. As we pass through a caustic in the direction of propagation, $\varrho+s$ changes sign and the correct phase shift of $+\pi/2$ is introduced naturally. Equation (6.8) is a valid high frequency approximation on either side of the caustic; however the field at the caustic must be found from separate considerations [6.6, 7].

Employing the Maxwell curl equation $\nabla X \boldsymbol{E} = -j\omega\mu\boldsymbol{H}$, it follows from (6.3) that the leading term in the asymptotic approximation for the magnetic field is

$$H \sim \hat{s} \times E/Z_c, \tag{6.9}$$

where $Z_c = \sqrt{\mu/\varepsilon}$ is the characteristic impedance of the medium, and E is given by (6.8).

6.1.2. Reflection

Geometrical optics provides a high-frequency approximation for the incident, reflected and refracted fields. Let us find the geometrical optics field

$$E^{rg}(s) = e^{-jk_0\psi^r(s)} E_0^r(s) \tag{6.10}$$

reflected from the point Q_R on a perfectly-conducting smooth curved surface S; the distance between Q_R and the field point on the reflected ray is denoted by s. The outward directed unit normal vector at Q_R is $\hat{n}$, and $\hat{s}^i$ and $\hat{s}^r$ are unit vectors in the directions of incidence and reflection, respectively, as shown in Fig. 6.2.

From the boundary condition on the total electric field at Q_R on S,

$$\hat{n} \times (E^i + E^r) = 0, \tag{6.11}$$

it can be shown that

$$E_0^r(Q_R) = E_0^i(Q_R) \cdot \underset{\sim}{R} = E_0^i(Q_R) \cdot [\hat{e}_{\|}^i \hat{e}_{\|}^r - \hat{e}_{\perp} \hat{e}_{\perp}] \tag{6.12}$$

where $\hat{e}_{\perp}$ is a unit vector perpendicular to the plane of incidence, and $\hat{e}_{\|}^i$, $\hat{e}_{\|}^r$ are unit vectors parallel to the plane of incidence so that

$$\hat{e}_{\|} = \hat{e}_{\perp} \times \hat{s}. \tag{6.13}$$

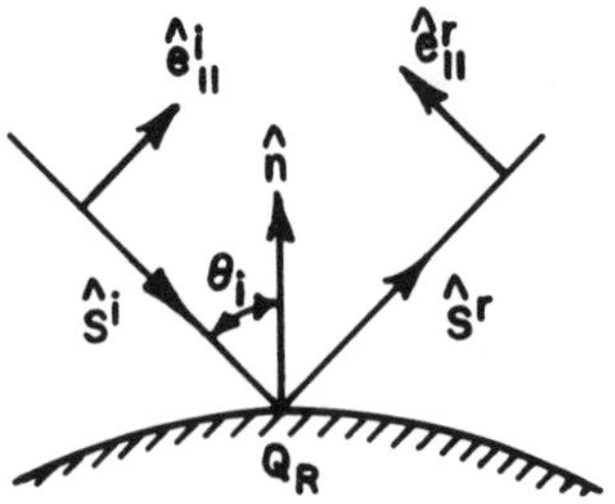

Fig. 6.2. Reflection at a curved surface

In matrix notation the reflection coefficient has a form familiar for the reflection of a plane electromagnetic wave from a plane, perfectly-conducting surface, namely

$$R = \begin{bmatrix} 1 & 0 \\ 0 & -1 \end{bmatrix}. \tag{6.14}$$

This is not surprising if one considers the local nature of high-frequency reflection, i.e., the phenomenon for the most part depends on the geometry of the problem in the immediate neighborhood of Q_R. Note that the incident and reflected fields must be phase matched on S to satisfy (6.11), i.e.,

$$\psi^{i}(Q_R) = \psi^{r}(Q_R). \tag{6.15}$$

The above equality leads to the law of reflection, and it is also employed to obtain (6.12).

The geometrical optics reflected field is

$$\boldsymbol{E}^{rg}(s) = \boldsymbol{E}^{ig}(Q_R) \cdot \underset{\sim}{\boldsymbol{R}} \sqrt{\frac{\varrho_1^r \varrho_2^r}{(\varrho_1^r + s)(\varrho_2^r + s)}}\, e^{-jks} \tag{6.16}$$

in which ϱ_1^r, ϱ_2^r are the principal radii of curvature of the reflected wave-front at Q_R. It can be shown that

$$\frac{1}{\varrho_1^r} = \frac{1}{2}\left(\frac{1}{\varrho_1^i} + \frac{1}{\varrho_2^i}\right) + \frac{1}{f_1} \tag{6.17}$$

$$\frac{1}{\varrho_2^r} = \frac{1}{2}\left(\frac{1}{\varrho_1^i} + \frac{1}{\varrho_2^i}\right) + \frac{1}{f_2}. \tag{6.18}$$

The above equations are reminiscent of the simple lens and mirror formulas of elementary physics; this is particularly true of an incident spherical wave, where $\varrho_1^i = \varrho_2^i = s'$. Expressions for f_1 and f_2 are given in [6.8]. For an incident spherical wave,

$$\frac{1}{f_{\frac{1}{2}}} = \frac{1}{\cos\theta_i}\left(\frac{\sin^2\theta_2}{a_1} + \frac{\sin^2\theta_1}{a_2}\right) \pm \sqrt{\frac{1}{\cos^2\theta_i}\left(\frac{\sin^2\theta_2}{a_1} + \frac{\sin^2\theta_1}{a_2}\right)^2 - \frac{4}{a_1 a_2}} \tag{6.19}$$

in which θ_1 and θ_2 are the angles between $\hat{s}^i$ and the principal directions associated with the principal radii of curvature of the surface a_1 and a_2, respectively. In the case of plane wave illumination it follows from (6.17–19),

$$\sqrt{\varrho_1^r \varrho_2^r} = \sqrt{a_1 a_2}/2 \tag{6.20}$$

which is useful in calculating the far-zone reflected field.

If a_1 or a_2 become infinite, as in the case of a flat plate or cylindrical scatterers, it is evident that geometrical optics fails. Geometrical optics approximates the scattered field only in the direction of specular reflection, as determined by the law of reflection.

In principle the geometrical-optics approximation can be improved by finding the higher-order terms $\boldsymbol{E}_1$, $\boldsymbol{E}_2, \dots$ in the Luneberg-Kline expansion. Luneberg-Kline expansions for fields reflected from cylinders, spheres and other curved surfaces with simple geometries are given in [6.9]. These terms improve the high-frequency approximation of the scattered field if the specular point is well away from shadow boundaries, edges or other surface discontinuities; however, it is noted that they tend to become singular as the specular point approaches close to a shadow boundary on the surface. Furthermore, these terms do not describe the diffracted field which penetrates into the shadow region, nor do they correct the discontinuities in the geometrical-optics field at shadow and reflection boundaries. An examination of available asymptotic solutions for diffracted fields reveals that they contain fractional powers of ω. Furthermore, one notes that caustics of the diffracted field are located at the boundary surface. From these considerations it is evident that the Luneberg-Kline series cannot be used to treat diffraction. At the present time additional postulates are required to introduce the high-frequency diffracted field; these are given in the next subsection.

It should ne noted, however, that for ω sufficiently large the geometrical-optics field may require no correction, i.e., the scattering phenomenon is entirely dominated by geometrical optics. This is the case for backscatter from smooth curved surfaces with radii of curvature very large in terms of a wavelength.

6.1.3. Diffraction

To overcome the limitations of the geometrical-optics field pointed out at the end of the last subsection, it is necessary to introduce an additional field, the diffracted field. KELLER [6.10–12] has shown how the diffracted field may be included in the high-frequency solution as an extension of geometrical optics. The postulates of Keller's theory, commonly referred to as the geometrical theory of diffraction (GTD), are summarized as follows.

(1) The diffracted field propagates along rays which are determined by a *generalization of Fermat's principle* to include points on the boundary surface in the ray trajectory.

2) Diffraction like reflection and transmission is a *local phenomenon* at high frequencies, i.e., it depends only on the nature of the boundary surface and the incident field in the immediate neighborhood of the point of diffraction.

3) The diffracted wave propagates along its ray so that

a) power is conserved in a tube (or strip of rays),

b) the phase delay along the ray path equals the product of the wave number of the medium and the distance.

The diffracted rays which pass through a given field point are found from the generalized Fermat's principle. The notion that points on the boundary surface may be included in the ray trajectory is not new. Imposing the condition that a point on a smooth curved surface be included in the ray path between the source and observation point is a time-honored method for deducing the reflected ray and the law of reflection. It seems reasonable to extend the class of such points as KELLER has done. Diffracted rays are initiated at points on the boundary surface where the incident geometrical-optics field is discontinuous[1], i.e., at points on the surface where there is a shadow or reflection boundary of the incident field. Examples of such points are edges, vertices and points at which the incident ray is tangent to a smooth, curved surface. The diffracted rays like the geometrical-optics rays follow paths which make the optical distance between the source point and the field point an extremum, usually a minimum. Thus the portion of the ray path which traverses a homogeneous medium is a straight line, and if a segment of the ray path lies on a smooth surface, it is a surface extremum or geodesic.

For points away from the diffracting surface, Postulate 3 of Keller's theory actually follows from his first two postulates. Consider a normal congruence of rays emanating from a point of diffraction on the radiating surface. The high-frequency diffracted field at P, see Fig. 6.3, may be found by asymptotically approximating its integral representation

$$\boldsymbol{E}^{\mathrm{d}}(s) = \int \boldsymbol{F}\, \mathrm{e}^{-jkr}\, da\,. \tag{6.21}$$

[1] The incident field may be a diffracted field; a discontinuity of the diffracted field on the boundary surface initiates a higher-order diffracted ray.

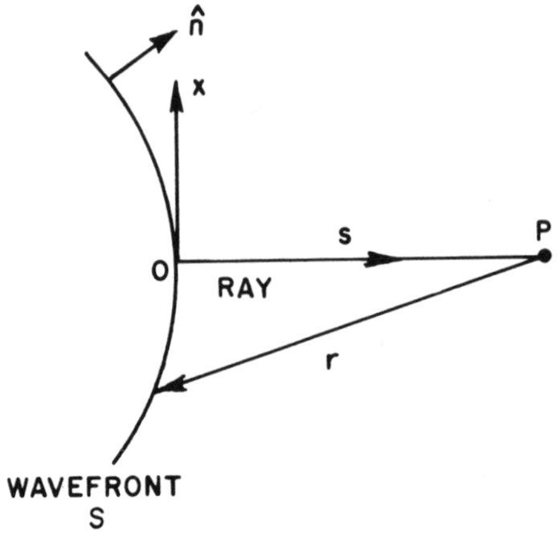

Fig. 6.3. Ray and wavefront geometry

The integral is taken over a wavefront associated with the diffracted rays and

$$\boldsymbol{F} = \frac{jkZ_c}{4\pi r}\left\{\hat{\boldsymbol{r}}\times[\hat{\boldsymbol{r}}\times(\hat{\boldsymbol{n}}\times\boldsymbol{H}^{\mathrm{d}})] - \frac{1}{Z_c}\hat{\boldsymbol{r}}\times(\boldsymbol{E}^{\mathrm{d}}\times\hat{\boldsymbol{n}})\right\} \tag{6.22}$$

in which $\hat{\boldsymbol{n}}$ is the unit normal vector to the wavefront and the other quantities are defined in Fig. 6.3. Then employing (6.9) together with the approximations

$$r \approx s + \frac{x^2}{2}\left(\frac{1}{s}+\frac{1}{\varrho_1}\right) + \frac{y^2}{2}\left(\frac{1}{s}+\frac{1}{\varrho_2}\right)$$

and

$$da \approx dx\,dy/\hat{\boldsymbol{n}}\cdot\hat{\boldsymbol{s}}$$

in which x, y are rectangular coordinates perpendicular to s at 0, one obtains

$$\boldsymbol{E}^{\mathrm{d}}(s) \sim \boldsymbol{E}^{\mathrm{d}}(0)\sqrt{\frac{\varrho_1\varrho_2}{(\varrho_1+s)(\varrho_2+s)}}\,\mathrm{e}^{-jks} \tag{6.23}$$

by using the method of stationary phase [6.13, 14]. Equation (6.23) has the same form as (6.8), which is not an unexpected result, because this development also can be applied to the incident and reflected fields of geometrical optics; however, it does not yield higher-order terms as does the Luneberg-Kline expansion.

In calculating the diffracted field it is convenient to choose the point of diffraction on the boundary surface as the reference point 0. However, this point of diffraction is also a caustic of the diffracted ray. First consider the case where the caustic is at an edge or forms a line on a smooth convex surface from which rays shed tangentially. Either ϱ_1 or ϱ_2 denoted by ϱ' vanishes; however, $\boldsymbol{E}^{\mathrm{d}}(s)$ must be independent of the location of the reference point; hence it follows from (6.23) that

$$\lim_{\varrho' \to 0} \boldsymbol{E}^{\mathrm{d}}(0)\sqrt{\varrho'} = \boldsymbol{C} \tag{6.24}$$

exists, so that

$$\boldsymbol{E}^{\mathrm{d}}(s) = \boldsymbol{C}\sqrt{\frac{\varrho}{s(\varrho + s)}}\,\mathrm{e}^{-jks} \tag{6.25}$$

in which ϱ is the distance between the caustic on the boundary surface (the point of diffraction) and the second caustic of the diffracted ray, which is away from this surface. Thus the diffracted rays, like the geometrical-optics rays form an astigmatic tube, as shown in Fig. 6.1 with either the caustic 1–2 or 3–4 at the point of diffraction on the boundary surface. The caustic distance ϱ may be determined by differential geometry; an expression for ϱ will be given later.

For a two-dimensional problem we note that $\varrho = \infty$, so (6.25) reduces to

$$\boldsymbol{E}^{\mathrm{d}}(s) = \boldsymbol{C}\,\frac{\mathrm{e}^{-jks}}{\sqrt{s}}\,. \tag{6.26}$$

Next let us consider the diffraction from a vertex or corner, where the diffracted rays emanate from a point caustic at the tip. In this case $\varrho_1 = \varrho_2 = \varrho'$, and again since $\boldsymbol{E}^{\mathrm{d}}(s)$ must be independent of the reference point $s = 0$, it follows from (6.23) that

$$\lim_{\varrho' \to 0} \boldsymbol{E}^{\mathrm{d}}(0)\,\varrho' = \boldsymbol{B} \tag{6.27}$$

exists, and so for vertex or corner diffraction

$$\boldsymbol{E}^{\mathrm{d}}(s) = \boldsymbol{B}\,\frac{\mathrm{e}^{-jks}}{s}\,. \tag{6.28}$$

Diffraction is a local effect according to Postulate 2, and since we are dealing with a linear phenomenon, $\boldsymbol{C}$ and $\boldsymbol{B}$ must be proportional to the incident field at the point of diffraction, if the incident field is not rapidly varying there[2]. The constant of proportionality is referred to as a diffraction coefficient, and for electromagnetic fields it is a dyadic. It is convenient to determine this from the asymptotic solution of the simplest boundary value problem having the same local geometry as that near the point of diffraction. A problem of this type is referred to as a canonical problem; canonical problems are employed to determine the diffraction coefficients for edges, the diffraction coefficients and attenuation constants for smooth curved surfaces, and other parameters of the GTD.

6.2. Edge Diffraction

6.2.1. The Wedge

Let us consider the field radiated from a point source at 0 and observed at P in the presence of a perfectly-conducting wedge, as shown in Fig. 6.4a, where the rays are projected on a plane perpendicular to the edge at the point of diffraction Q_E. Applying the generalized Fermat's principle, the distance along the ray path $0Q_EP$ is a minimum and the law of edge diffraction

$$\hat{s}' \cdot \hat{e} = \hat{s} \cdot \hat{e} \tag{6.29}$$

results. Here $\hat{e}$ is a unit vector directed along the edge, and $\hat{s}'$ and $\hat{s}$ are unit vectors in the directions of incidence and diffraction, respectively. The above equation also follows from the requirement that the incident and diffracted fields be phase matched along the edge. If the incident ray strikes the edge obliquely, making an angle β_0' with the edge, as shown in Fig. 6.4b, the diffracted rays lie on the surface of a cone whose half angle is equal to β_0'. The position of the diffracted ray on this conical surface is given by the angle ϕ, and the direction of the ray incident on the edge, by the angles ϕ' and β_0'; these angles are defined in Fig. 6.4a and b. Equation (6.29) may be used to develop a computer search program to locate the points of edge diffraction.

From (6.25) and the discussion following, the expression for the electric field of the edge-diffracted ray is

$$\boldsymbol{E}^{\mathrm{d}}(s) = \boldsymbol{E}^{\mathrm{i}}(\mathrm{Q_E}) \cdot \underset{\sim}{\boldsymbol{D}}(\phi, \phi'; \beta_0') \sqrt{\frac{\varrho}{s(\varrho + s)}}\, \mathrm{e}^{-jks} \tag{6.30}$$

[2] If the incident field is rapidly varying at the point of diffraction, it may be possible to separate it into slowly-varying components for the purpose of calculating the diffracted field.

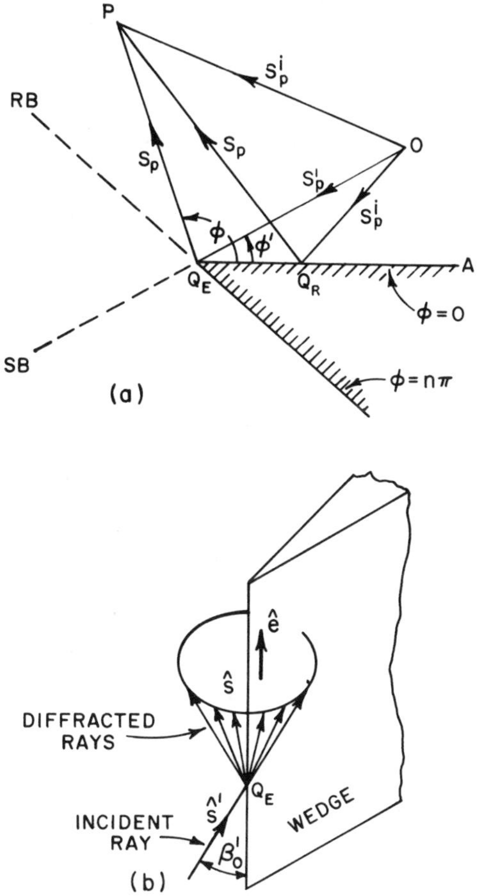

Fig. 6.4a and b. Reflection and diffraction from a wedge. The subscript p indicates that the ray path is shown projected on a plane perpendicular to the edge

in which $\underset{\sim}{D}(\phi, \phi'; \beta_0')$ is the dyadic edge-diffraction coefficient. Since the pertinent dimension in wedge diffraction is wavelength, it follows from dimensional considerations that the diffraction coefficient must vary as $k^{-1/2}$. The dyadic diffraction coefficient for a perfectly-conducting wedge has been obtained by KOUYOUMJIAN and PATHAK; their work is described in [6.8, 15] and will only be summarized here. As noted before, the dyadic diffraction coefficient can be found from the asymptotic solution of canonical problems, which in this case involve the illumination of the wedge by plane, cylindrical, conical and spherical waves. The solution of these canonical problems serves as a basis for deducing the dyadic diffraction coefficient for arbitrary wavefront illumination and for the more general case where there are curved edges and curved surfaces.

Let us introduce an edge-fixed plane of incidence containing the incident ray and the edge and a plane of diffraction containing the diffracted ray and the edge. The unit vectors $\hat{\boldsymbol{\phi}}'$ and $\hat{\boldsymbol{\phi}}$ are perpendicular to the edge-fixed plane of incidence and the plane of diffraction, respectively. The unit vectors $\hat{\boldsymbol{\beta}}_0'$ and $\hat{\boldsymbol{\beta}}_0$ are parallel to the edge-fixed plane of incidence and the plane of diffraction, respectively, and

$$\hat{\boldsymbol{\beta}}_0' = \hat{\boldsymbol{s}}' \times \hat{\boldsymbol{\phi}}', \qquad \hat{\boldsymbol{\beta}}_0 = \hat{\boldsymbol{s}} \times \hat{\boldsymbol{\phi}} .$$

Thus the coordinates of the diffracted ray (s, β_0, ϕ) are spherical coordinates and so are the coordinates of the incident ray (s', β_0', ϕ'),

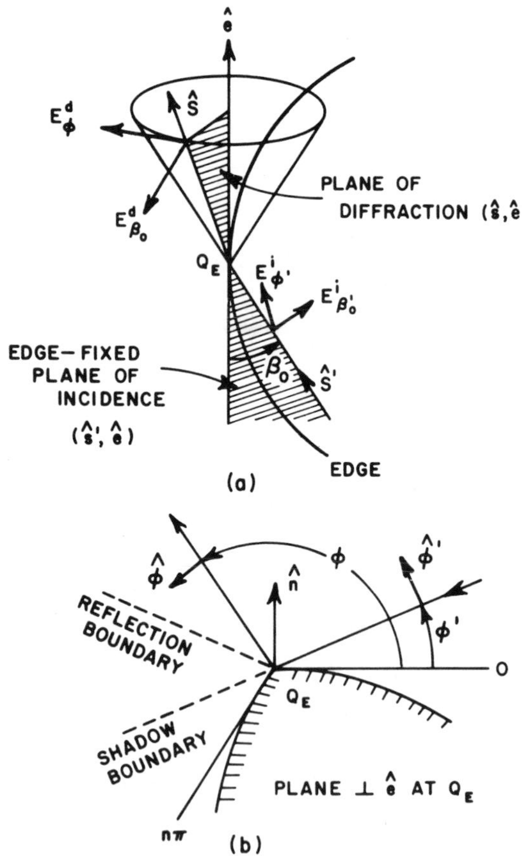

Fig. 6.5a and b. Diffraction by a curved edge

except that the incident (radial) unit vector points toward the origin Q_E. These ray-fixed coordinates and their unit vectors are shown in Figs. 6.4 and 6.5. Although the latter figure depicts curved edges and curved surfaces, it is still helpful in the present discussion (one may regard the wedge as just a special case of the curved edge structure).

For each type of edge illumination mentioned previously, it is shown in [6.15] that the dyadic diffraction coefficient can be represented simply as the sum of two dyads, if the ray-fixed coordinates mentioned in the preceding paragraph are used.

$$\underset{\sim}{D}(\phi, \phi'; \beta'_0) = -\hat{\boldsymbol{\beta}}'_0 \hat{\boldsymbol{\beta}}_0 D_s(\phi, \phi'; \beta'_0) - \hat{\boldsymbol{\phi}}' \hat{\boldsymbol{\phi}} D_h(\phi, \phi'; \beta'_0), \tag{6.31}$$

where D_s is the scalar diffraction coefficient for the acoustically soft (Dirichlet) boundary condition at the surface of the wedge, and D_h is the scalar diffraction coefficient for the acoustically hard (Neumann) boundary condition. This result shows the close connection between electromagnetics and acoustics at high frequencies. If the dyadic diffraction coefficient is expressed in an edge-fixed coordinate system, it is found to be the sum of seven dyads. In matrix notation this means that the diffraction coefficient is a 3×3 matrix with seven non-vanishing elements, instead of the 2×2 diagonal matrix which may be used to represent the diffraction coefficient in the ray-fixed coordinate system. In this sense the ray-fixed coordinate system is the natural coordinate system of the problem.

If the field point is not close to a shadow or reflection boundary and $\phi' \neq 0$ or $n\pi$, the scalar diffraction coefficients

$$D_{\substack{s\\h}}(\phi, \phi'; \beta'_0) = \frac{e^{-\frac{\pi}{4}} \sin \frac{\pi}{n}}{n\sqrt{2\pi k} \sin \beta'_0} \cdot \left[\frac{1}{\cos \frac{\pi}{n} - \cos\left(\frac{\phi - \phi'}{n}\right)} \mp \frac{1}{\cos \frac{\pi}{n} - \cos\left(\frac{\phi + \phi'}{n}\right)} \right] \tag{6.32}$$

for all four types of illumination, which is important because the diffraction coefficient should be independent of the edge illumination away from shadow and reflection boundaries. The wedge angle is $(2-n)\pi$, where the plane surfaces forming the wedge are $\phi = 0$ and $\phi = n\pi$, see Fig. 6.4a. This expression becomes singular as a shadow boundary (SB) or a reflection boundary (RB) is approached, which further aggravates the difficulties at these boundaries resulting from the discontinuities in the incident or reflected fields. The above scalar diffraction coefficients also have been given by KELLER [6.12]. The case of grazing incidence $\phi'=0$ or $n\pi$ will be considered later.

Expressions for the scalar diffraction coefficients which are valid at all points away from the edge (again excluding $\phi'=0$ or $n\pi$) are

$$D_{\substack{s\\h}}(\phi,\phi';\beta_0')=\frac{-e^{-j\frac{\pi}{4}}}{2n\sqrt{2\pi k}\sin\beta_0'}$$
$$\times\left[\cot\left(\frac{\pi+(\phi-\phi')}{2n}\right)F[kLa^{+}(\phi-\phi')]\right.$$
$$+\cot\left(\frac{\pi-(\phi-\phi')}{2n}\right)F[kLa^{-}(\phi-\phi')] \tag{6.33}$$
$$\mp\left\{\cot\left(\frac{\pi+(\phi+\phi')}{2n}\right)F[kLa^{+}(\phi+\phi')]\right.$$
$$\left.\left.+\cot\left(\frac{\pi-(\phi+\phi')}{2n}\right)F[kLa^{-}(\phi+\phi')]\right\}\right],$$

where

$$F(X)=2j|\sqrt{X}|\,e^{jX}\int\limits_{|\sqrt{X}|}^{\infty}e^{-j\tau^2}d\tau. \tag{6.34}$$

When X is small

$$F(X)\simeq\left[\sqrt{\pi X}-2Xe^{j\frac{\pi}{4}}-\tfrac{2}{3}X^2e^{-j\frac{\pi}{4}}\right]e^{j\left(\frac{\pi}{4}+X\right)}, \tag{6.35}$$

and when X is large

$$F(X)\sim\left(1+j\frac{1}{2X}-\frac{3}{4}\frac{1}{X^2}-j\frac{15}{8}\frac{1}{X^3}+\frac{75}{16}\frac{1}{X^4}\right). \tag{6.36}$$

If the arguments of the four transition functions in (6.33) exeed 10, the transition functions are approximately equal to one, and (6.33) reduces to (6.32).

The argument of the transition function $X=kLa^{\pm}(\phi\pm\phi')$ in which L is a distance parameter which will be given later. The large parameter in the asymptotic approximation is kL. Let $\phi\pm\phi'=\beta$, then

$$\alpha^{\pm}(\beta)=2\cos^2\left(\frac{2n\pi N^{\pm}-\beta}{2}\right) \tag{6.37}$$

in which $N^{\pm}$ are the integers which most nearly satisfy the equations

$$2\pi n N^{+} - \beta = \pi, \tag{6.38}$$

$$2\pi n N^{-} - \beta = -\pi. \tag{6.39}$$

Note that N^{+} and N^{-} each have two values.

$a^{\pm}(\beta)$ is a measure of the angular separation between the field point and a shadow or reflection boundary. The + and − superscripts are associated with the integers N^{+} and N^{-}, respectively, which are defined by (6.38) and (6.39). For exterior edge diffraction ($1 < n \leqq 2$), $N^{+} = 0$ or 1 and $N^{-} = -1, 0$ or 1.

At a shadow or reflection boundary one of the cotangent functions in the expression for D given by (6.33) becomes singular; the other three remain bounded. Even though the cotangent becomes singular, its product with the transition function can be shown to be bounded. The location of the boundary at which each cotangent becomes singular is presented in Table 6.1. Since discontinuity in the geometrical-optics field at a shadow or reflection boundary is compensated separately by one of the four terms in the diffraction coefficient, there is no problem in calculating the field when two boundaries are close to each other or coincide. The value of N^{+} or N^{-} at each boundary is included in Table 6.1 for convenience; these values remain unchanged in their respective transition regions unless kL is small.

The distance parameter L can be found by imposing the condition that the total field (the sum of the geometrical-optics field and the diffracted field) be continuous at shadow or reflection boundaries. One

Table 6.1. Behaviour of terms in the edge diffraction coefficient at shadow and reflection boundaries

	The cotangent is singular when	value of N at the boundary
$\cot\left(\frac{\pi + (\phi - \phi')}{2n}\right)$	$\phi = \phi' - \pi$, a SB surface $\phi = 0$ is shadowed	$N^{+} = 0$
$\cot\left(\frac{\pi - (\phi - \phi')}{2n}\right)$	$\phi = \phi' + \pi$, a SB surface $\phi = n\pi$ is shadowed	$N^{-} = 0$
$\cot\left(\frac{\pi + (\phi + \phi')}{2n}\right)$	$\phi = (2n - 1)\pi - \phi'$, a RB reflection from surface $\phi = n\pi$	$N^{+} = 1$
$\cot\left(\frac{\pi - (\phi + \phi')}{2n}\right)$	$\phi = \pi - \phi'$, a RB reflection from surface $\phi = 0$	$N^{-} = 0$

obtains

$$L = \frac{s(\varrho_e^i + s)\,\varrho_1^i\,\varrho_2^i \sin^2\beta_0'}{\varrho_e^i(\varrho_1^i + s)\,(\varrho_2^i + s)}, \tag{6.40}$$

where ϱ_1^i, ϱ_2^i are the principal radii of curvature of the incident wavefront at Q_E, and ϱ_e^i is the radius of curvature of the incident wavefront in the edge fixed plane of incidence. Note that $\varrho = \varrho_e^i$ for the wedge.

Grazing incidence, where $\phi' = 0$ or $n\pi$ must be considered separately. In this case $D_s = 0$, and the expression for D_h given by (6.32) or (6.33) is multiplied by a factor of 1/2. The need for the factor of 1/2 may be seen by considering grazing incidence to be the limit of oblique incidence. At grazing incidence the incident and reflected fields merge, so that one half the total field propagating along the face of the wedge toward the edge is the incident field and the other half is the reflected field. Nevertheless, in this case it is clearly more convenient to regard the total field as the "incident" field. $D_s = 0$ implies that $E_{\beta_0}^d$ vanishes; however, as pointed out by KARP and KELLER [6.16], a higher-order term then becomes significant where $E_{\beta_0}^d$ is proportional to the normal derivative of $E_{\beta_0'}^i(Q_E)$. It can be shown for $\phi' = 0$ that

$$E_{\beta_0}^d(s) = \frac{1}{2}\,\frac{\partial E_{\beta_0'}^i(Q_E)}{\partial n} \cdot \frac{1}{jk\sin\beta_0}\,\frac{\partial}{\partial\phi'}\,D_s\bigg|_{\phi'=0} \sqrt{\frac{\varrho}{s(\varrho + s)}}\,e^{-jks}, \tag{6.41}$$

where D_s is given by (6.33). However, unlike (Ref. [16], Eq. (8)), the above expression may be used in the transition region adjacent to the shadow boundary at $\phi = \pi$. It was found to give accurate values of the diffracted field in the case of an infinitesimal slot (magnetic dipole) perpendicular to the edge of the wedge when the slot is only a quarter wavelength from the edge.

6.2.2. The General Edge Configuration

In the general case the surfaces forming the edge may be convex, concave or plane. Our solution is based of Keller's method of the canonical problem. The justification of the method is that high-frequency diffraction like high-frequency reflection is a local phenomenon, and locally one can approximate the curved edge geometry by a wedge, where the straight edge of the wedge is tangent to the curved edge at the point of incidence Q_E in Fig. 6.5 and its plane surfaces are tangent to the surfaces forming the curved edge. With these assumptions, the results for wedge diffraction can be applied directly to the curved edge problem. As we have just noted, there is an equivalent wedge associated with every curved edge structure, and so in generalizing the solution of the wedge, it is only necessary to modify the expressions for the distance parameter L, which appear in the arguments of the transition functions.

Thus the form of the dyadic diffraction coefficient is given by (6.31); the unit vectors and coordinates are shown in Fig. 6.5. It remains to determine the scalar diffraction coefficients. Outside the transition regions; these are given by (6.32).

The calculation of the caustic distance ϱ in (6.30) is not a trivial matter for curved edge diffraction. Employing differential geometry, it is shown in [6.8, 15] that

$$\frac{1}{\varrho}=\frac{1}{\varrho_e^1}-\frac{\hat{\boldsymbol{n}}_e\cdot(\hat{s}'-\hat{s})}{a\sin^2\beta_0'} \tag{6.42}$$

in which ϱ_e^i is the radius of curvature of the incident wavefront in the edge-fixed plane of incidence, which contains $\hat{s}'$ and $\hat{e}$ the unit vector tangent to the edge at Q_E; $\hat{\boldsymbol{n}}_e$ is the unit vector normal to the edge at Q_E and directed away from the center of curvature; $\hat{s}'$ and $\hat{s}$ are unit vectors in the directions of incidence and diffraction, respectively, see Fig. 6.5a; a is the radius of curvature of the edge at Q_E, $a>0$.

It is interesting to note that (6.42) like (6.17) and (6.18) has the same form as the elementary lens equation; here ϱ_e^1 and ϱ correspond to the object and image distances, respectively.

As in the case of the wedge, the arguments of the transition functions are determined by imposing the condition that the total field be continuous at the shadow and reflection boundaries. It is found that

$$\begin{aligned}
D_{\substack{s\\h}}(\phi,\phi';\beta_0') = & \frac{-\mathrm{e}^{-j\frac{\pi}{4}}}{2n\sqrt{2\pi k}\sin\beta_0'} \\
& \times\left[\cot\left(\frac{\pi+(\phi-\phi')}{2n}\right)F[kL^i a^+(\phi-\phi')]\right. \\
& +\cot\left(\frac{\pi-(\phi-\phi')}{2n}\right)F[kL^i a^-(\phi-\phi')] \\
& \mp\left\{\cot\left(\frac{\pi+(\phi+\phi')}{2n}\right)F[kL^{ro} a^+(\phi+\phi')]\right. \\
& \left.\left.+\cot\left(\frac{\pi-(\phi+\phi')}{2n}\right)F[kL^{rn} a^-(\phi+\phi')]\right\}\right],
\end{aligned} \tag{6.43}$$

$F(X)$, $a^{\pm}(\beta)$, $N^{\pm}$ being defined as before, and

$$L^{\mathrm{i}} = \frac{s(\varrho_{\mathrm{e}}^{\mathrm{i}} + s)\,\varrho_2^{\mathrm{i}}\varrho_1^{\mathrm{i}}\sin^2\beta_0'}{\varrho_{\mathrm{e}}^{\mathrm{i}}(\varrho_1^{\mathrm{i}} + s)\,(\varrho_2^{\mathrm{i}} + s)}, \tag{6.44}$$

$$L^{\mathrm{r}} = \frac{s(\varrho^{\mathrm{r}} + s)\,\varrho_2^{\mathrm{r}}\varrho_1^{\mathrm{r}}\sin^2\beta_0'}{\varrho^{\mathrm{r}}(\varrho_1^{\mathrm{r}} + s)\,(\varrho_2^{\mathrm{r}} + s)}, \tag{6.45}$$

where ϱ_1^{r} and ϱ_2^{r} are the principal radii of curvature of the reflected wavefront at Q_{E}, and ϱ^{r} is the distance between the caustics of the diffracted ray in the direction of reflection. It may be found from (6.42) with $\hat{\boldsymbol{s}} = \hat{\boldsymbol{s}}' - 2(\hat{\boldsymbol{n}} \cdot \hat{\boldsymbol{s}}')\hat{\boldsymbol{n}}$. The additional superscripts o and n on L in (6.43) denote that the radii of curvature (and caustic distance ϱ) are calculated at the reflection boundaries $\pi - \phi'$ and $(2n-1)\pi - \phi'$, respectively. In the far-zone where $s \gg \varrho$ and the principal radii of curvature ϱ_1 and ϱ_2 of the incident and reflected wavefronts at Q_{E}, (6.44) and (6.45) simplify to

$$L = \frac{\varrho_1\varrho_2\sin^2\beta_0'}{\varrho}; \tag{6.46}$$

the appropriate superscripts are omitted here for the sake of simplicity.

Usually no more than one of the four transition functions is significantly different from one; furthermore the nature of the curved edge approximation is such that the first two terms within the brackets of (6.43) can be combined [6.8] to give

$$\frac{-2\sin\dfrac{\pi}{n}\,F\left[2kL^{\mathrm{i}}\cos^2\left(\dfrac{\phi-\phi'}{2}\right)\right]}{\cos\dfrac{\pi}{n} - \cos\left(\dfrac{\phi-\phi'}{n}\right)},$$

which considerably simplifies the calculation of the scalar diffraction coefficients.

In summary to calculate the diffraction from a curved edge, the scalar diffraction coefficients from (6.43) are substituted into (6.31), and the resulting dyadic diffraction coefficient is substituted into (6.30). The caustic distance ϱ is calculated from (6.42). In matrix notation

$$\begin{bmatrix} E_{\beta_0}^{\mathrm{d}} \\ E_{\phi}^{\mathrm{d}} \end{bmatrix} = \begin{bmatrix} -D_{\mathrm{s}} & 0 \\ 0 & -D_{\mathrm{h}} \end{bmatrix} \begin{bmatrix} E_{\beta_0'}^{\mathrm{i}} \\ E_{\phi'}^{\mathrm{i}} \end{bmatrix} \sqrt{\frac{\varrho}{s(\varrho+s)}}\,\mathrm{e}^{-jks}. \tag{6.47}$$

The present treatment does not include the modification of the edge diffracted field which occurs when either the incident or diffracted ray grazes the surface. The angle between these rays and the surface should exceed $(ka_i)^{-1/3}$, where a_i is the radius of curvature of the surface at Q_E in the direction of the incident (or diffracted) ray. Also the present treatment does not include the effect of surface rays excited at the edge.

6.2.3. Higher-Order Edges

The preceding discussion has been restricted to ordinary edges where the unit normal vector to the surface is discontinuous. However, in the case of higher-order edges, where some j-th derivative of the surface has a jump discontinuity (while all lower derivatives are continuous), it has been shown [6.8] that the dyadic diffraction coefficient has the same form as in (6.31). Also the dyadic diffraction coefficient for the scattering from thin, curved wires has this form too.

Recently, KELLER and KAMINETZKY [6.17] and SENIOR [6.18] have obtained expressions for the scalar diffraction coefficients in the case of diffraction by an edge formed by a discontinuity in surface curvature and SENIOR has given the dyadic (or matrix) diffraction coefficient in an edge-fixed coordinate system. When transformed to the ray-fixed coordinate system, Senior's expression for the diffracted field reduces to the form in (6.30) with $\underset{\sim}{\boldsymbol{D}}$ given by (6.31). KELLER and KAMINETZKY [6.17] also have given expressions for the scalar diffraction coefficients in the case of higher-order edges.

6.3. Diffraction by a Vertex

The vertex or corner is a point caustic of the diffracted rays, from (6.28) and the discussion following, the expression for the electric field of the vertex diffracted ray is

$$\boldsymbol{E}^{\mathrm{d}}(s)=\boldsymbol{E}^{\mathrm{i}}(\mathrm{Q_v})\cdot\underset{\sim}{\boldsymbol{D}}(\theta,\phi;\theta',\phi')\,\frac{\mathrm{e}^{-jks}}{s}. \tag{6.48}$$

From dimensional considerations similar to those applied to the wedge, this diffraction coefficient must vary as k^{-1}, which means that outside its transitions regions the vertex-diffracted field is in general significantly weaker than the edge-diffracted field. Very little work has been done on the high-frequency diffraction by vertices; it is a complicated, difficult subject, and there are a variety of such geometries to consider. Some results for blunt vertices may be obtained from [6.17], and the diffraction coefficient for the vertex of a cone in some special cases may be obtained from [6.19, 20].

6.4. Surface Diffraction

When an incident ray strikes a smooth, curved perfectly-conducting surface at grazing incidence, i.e., at the shadow boundary, a part of its energy is diffracted into the shadow region. Let us consider the field radiated by the source 0 and observed at P in the shadow region, as shown in Fig. 6.6. Applying the generalized Fermat's principle, the distance $0Q_1Q_2P$ is the shortest distance between 0 and P which does not penetrate the surface. In detail, a ray incident on the shadow boundary

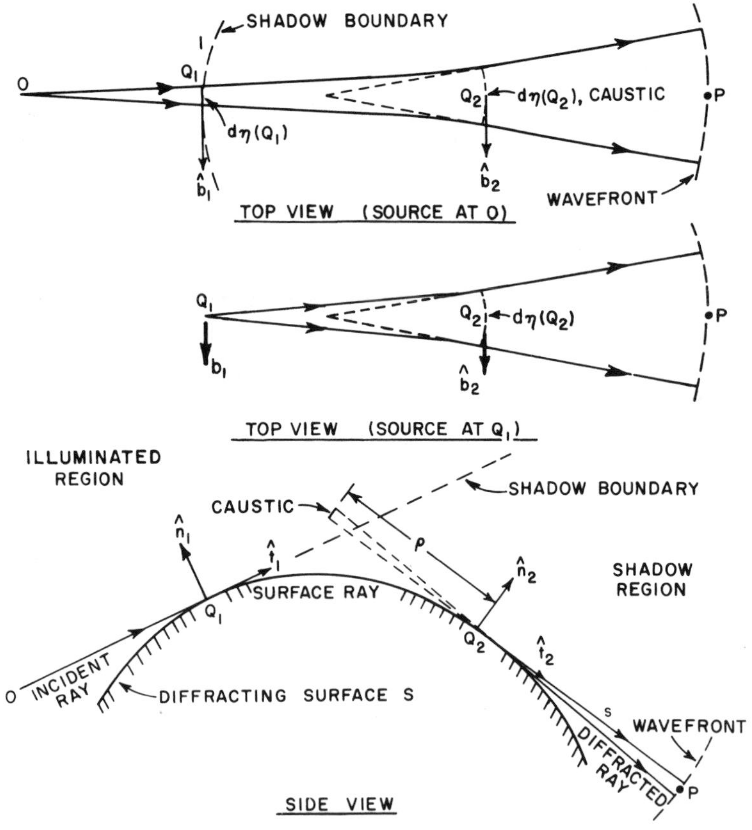

Fig. 6.6. Diffraction by a smooth convex surface

at Q_1 divides; one part of the incident energy continues straight on as predicted by geometrical optics and a second part follows the surface S into the shadow region as a surface ray, which sheds diffracted rays tangentially as it propagates. It follows from this extension of Fermat's principle that the incident and diffracted rays are tangent to S and to the surface ray at Q_1 and Q_2, respectively, and that the surface ray is the shortest distance between Q_1 and Q_2 on S, i.e., the surface ray is a geodesic curve. The former statement is referred to as the law of surface diffraction, and it also may be deduced from the requirement that the incident and diffracted fields are phase matched to the surface ray field at Q_1 and Q_2. At Q_1 let $\hat{\boldsymbol{t}}_1$ be the unit vector in direction of incidence, $\hat{\boldsymbol{n}}_1$ be the unit vector normal to S and $\hat{\boldsymbol{b}}_1 = \hat{\boldsymbol{t}}_1 \times \hat{\boldsymbol{n}}_1$; at Q_2 let a similar set of unit vectors be defined with $\hat{\boldsymbol{t}}_2$ in the direction of the diffracted ray.

A second configuration of interest occurs when the source is positioned on the surface, say at Q_1. This configuration is relevant to the radiation from an aperture in S, where the equivalent source is an infinitesimal magnetic current moment (magnetic dipole)

$$d\boldsymbol{p}_{\mathrm{m}}(Q_1) = \boldsymbol{E}(Q_1) \times \hat{\boldsymbol{n}}_1 \, da \tag{6.49}$$

in which $\boldsymbol{E}$ is the aperture electric field, and da is an area element of the aperture. Another type of source which may be positioned at Q_1 is the normally-directed electric current dipole. According to the generalized Fermat's principle, the ray trajectory from these sources to P is the curve $Q_1 Q_2 P$ mentioned previously.

The discussion to follow is devoted to methods of calculating the field in the shadow and transition regions of a convex, perfectly-conducting surface. Deep in the illuminated region the field directly radiated from the source is found by geometrical optics. Expressions for this field are well known and will not be repeated here. When the source is not on the surface the reflected field may be calculated from (6.14), (6.16)–(6.19).

6.4.1. The Shadow Region

From Fig. 6.6 it is seen that Q_2 is a caustic of the diffracted field. There is a second caustic at a distance ϱ from this caustic. As noted earlier, the diffracted field at P is given by (6.25), where it is convenient to let

$$\boldsymbol{C} = \hat{\boldsymbol{n}}_2 C_n + \hat{\boldsymbol{b}}_2 C_b \tag{6.50}$$

in this case. The C_n, C_b are proportional to the surface ray field incident at Q_2; however, as in the case of edge diffraction, the precise relationship (like much of the development to follow) is deduced from the asymptotic

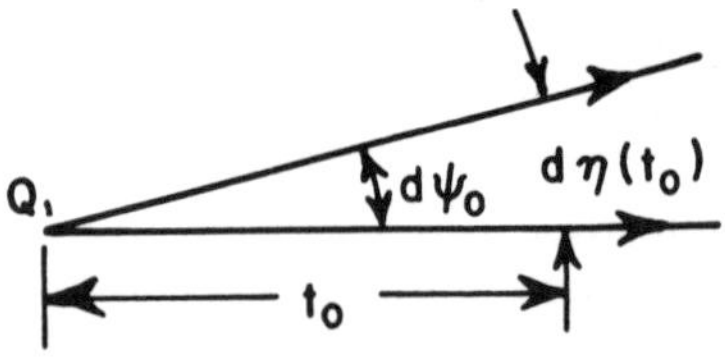

Fig. 6.7. Surface ray configuration close to a point source

solution of certain canonical problems, which will be described later. To simplify the discussion, it is assumed that the surface rays are torsionless, i.e., $\hat{\boldsymbol{b}}$ does not change direction along the surface ray.

From the canonical problems it is found that the surface ray field is composed of infinitely many modes which propagate independently of each other along a torsionless path. Let the field associated with one of these modes be

$$a(t) = A(t)\,e^{j(\phi_0 - kt)}, \tag{6.51}$$

where t is the distance along the surface ray measured from Q_1, ϕ_0 is the phase at Q_1, and initially one assumes that $A(t)$ is real.

The surface ray sheds rays tangentially as it propagates along a geodesic on the curved surface; hence energy is continuously lost from the surface ray field, and the field of each mode is attenuated. In addition, it is assumed that the energy flux between adjacent surface rays is conserved. This may be expressed by

$$d/dt(A^2\,d\eta) = -2\alpha(A^2\,d\eta), \tag{6.52}$$

where α is the attenuation constant for the surface ray mode in question. The above equation is readily integrated between t_0 and t to give

$$a(t) = a(t_0)\sqrt{\frac{d\eta(t_0)}{d\eta(t)}}\,\exp\left\{-\left[jk(t-t_0) + \int_{t_0}^{t} \alpha(t')\,dt'\right]\right\}. \tag{6.53}$$

The attenuation constant is a function of t' because it depends on the local curvature of the surface.

Equation (6.53) must be modified when there is a caustic due to a point source on the surface at Q_1, where $t=0$. For t_0 small $d\eta(t_0) = t_0\,d\psi_0$ is the angle between adjacent surface rays, see Fig. 6.7. Moreover, $a(t)$ must be independent of t_0; hence $\lim a(t_0)\sqrt{t_0}$ exists as $t_0 \to 0$ and we define it to be K. It follows then that

$$a(t)=K\sqrt{\frac{d\psi_0}{d\eta}}\exp\left\{-\left[jkt+\int_0^t \alpha(t')\,dt'\right]\right\}. \tag{6.54}$$

The constant K is proportional to the strength of the source at Q_1.

For the source at 0 removed from the boundary surface, the incident field at Q_1 may be resolved into normal and binormal components, which induce hard (h) and soft (s) type surface ray modes, respectively. The normal derivative of the field at S vanishes for the hard boundary condition[3], and the field at S vanishes for the soft boundary condition. Thus

$$\hat{\boldsymbol{n}}_1\cdot\boldsymbol{E}^{\mathrm{i}}(1)\,D_p^{\mathrm{h}}(1)=a_p^{\mathrm{h}}(1)\,, \tag{6.55a}$$

$$\hat{\boldsymbol{b}}_1\cdot\boldsymbol{E}^{\mathrm{i}}(1)\,D_p^{\mathrm{s}}(1)=a_p^{\mathrm{s}}(1)\,, \tag{6.55b}$$

where the constant of proportionality D is the surface diffraction coefficient, the superscript h(s) denotes a quantity associated with the hard (soft) boundary condition, the subscript p denotes the pth surface ray mode and Q_1 and Q_2 are replaced by 1 and 2 for the sake of notational economy.

At Q_2 the components of $\boldsymbol{C}$ are linearly related to the surface ray field:

$$\sum_p a_p^{\mathrm{h}}(2)\,D_p^{\mathrm{h}}(2)=C_n\,, \tag{6.56a}$$

$$\sum_p a_p^{\mathrm{s}}(2)\,D_p^{\mathrm{s}}(2)=C_b\,. \tag{6.56b}$$

Now substituting (6.56) into (6.50) and (6.25), noting that $a_p(1)$ and $a_p(2)$ are related by (6.53) and employing (6.55), one obtains

$$\boldsymbol{E}^{\mathrm{d}}(P)=\boldsymbol{E}^{\mathrm{i}}(1)\cdot[\hat{\boldsymbol{n}}_1\,\hat{\boldsymbol{n}}_2\,F+\hat{\boldsymbol{b}}_1\,\hat{\boldsymbol{b}}_2\,G]\sqrt{\frac{\varrho}{s(\varrho+s)}}\,\mathrm{e}^{-jks} \tag{6.57}$$

in which

$$F=\mathrm{e}^{-jkt}\sqrt{\frac{d\eta_1}{d\eta_2}}\sum_{p=1}^{\infty}D_p^{\mathrm{h}}(1)\,D_p^{\mathrm{h}}(2)\;T_p^{\mathrm{h}} \tag{6.58}$$

---

[3] This approximation of the hard boundary condition for electromagnetic waves is adequate for the present discussion. A more complete treatment is given later.

and G has the same form as F except that the superscript h is replaced by s. Here t is the distance between Q_1 and Q_2 along the surface ray, and we have set

$$\exp\left[-\int_0^t \alpha_p^{\mathrm{h}}(t')\,dt'\right] = T_p^{\mathrm{h}} \tag{6.59}$$

for the sake of notational brevity. If the reciprocity principle is to be satisfied, $D_p(1)$ must have the same functional dependence as $D_p(2)$, i.e., if a source at 0 is to produce the same field at P as a source at P produces at 0. It is apparent that the bracketed quantity in (6.57) serves as a generalized diffraction coefficient for the convex surface, analogous to (6.31) for the edge.

If S is a closed surface, a surface ray initiated at Q_1 may encircle S an infinite number of times. The length of the surface ray path for the lth encirclement is $t + lT$ with T the length of the closed path. These multiply-encircling rays can be summed to contribute

$$1 - \exp\left\{-\left(jkT + \int_0^T \alpha_p(t')\,dt'\right)\right\}$$

to the denominator of the expression for the diffracted field.

If the source is on the surface at Q_1, one still employs (6.56), (6.50), and (6.25); however in this case $a_p(2)$ is related to $a_p(1)$ by (6.54), and at Q_1

$$\frac{-jk}{4\pi}\,d\boldsymbol{p}_m \cdot \hat{\boldsymbol{b}}_1\, L_p^{\mathrm{h}} = K_p^{\mathrm{h}}, \tag{6.60a}$$

$$\frac{-jk}{4\pi}\,d\boldsymbol{p}_m \cdot \hat{\boldsymbol{t}}_1\, L_p^{\mathrm{s}} = K_p^{\mathrm{s}}, \tag{6.60b}$$

where the constants of proportionality are referred to as launching coefficients. It follows then that

$$d\boldsymbol{E}^{\mathrm{d}}(P) = d\boldsymbol{p}_m \cdot [\hat{\boldsymbol{b}}_1 \hat{\boldsymbol{n}}_2 F_s + \hat{\boldsymbol{t}}_1 \hat{\boldsymbol{b}}_2 G_s] \sqrt{\frac{\varrho}{s(\varrho + s)}}\, \mathrm{e}^{-jks} \tag{6.61}$$

in which

$$F_s = \frac{-jk\mathrm{e}^{-jkt}}{4\pi} \sqrt{\frac{1}{\varrho}\frac{d\psi_1}{d\psi_2}} \sum_{p=1}^{\infty} L_p^{\mathrm{h}}(1)\, D_p^{\mathrm{h}}(2)\, T_p^{\mathrm{h}} \tag{6.62}$$

and G_s has the same form as F_s except that the superscript h is replaced by s. In (6.54) $d\psi_0$ has been replaced $d\psi_1$, and $d\eta = d\eta_2 = \varrho\, d\psi_2$. $\boldsymbol{E}^{\mathrm{d}}(P)$ is calculated by integrating over the aperture in question.

At first glance it may seem that the field on S can be calculated directly from the surface ray field. However, the surface ray field is not a physically observable field; as a matter of fact, it does not have the dimensions of an electric or magnetic field, as will be seen when the expressions for D_p are given. Thus in (6.56) the surface ray field merely serves as a transfer function between the incident field at Q_1 and the diffracted field at Q_2. However, both the surface ray field and the field on the surface vary with respect to t in the same manner, which makes it possible to calculate the magnetic field at Q_2 on the perfectly-conducting surface by introducing attachment coefficients A_p in the place of diffraction coefficients at Q_2.

Thus for a source at 0

$$\boldsymbol{H}(\mathrm{Q}_2) = Y_{\mathrm{c}} \boldsymbol{E}^{\mathrm{i}} \cdot [\hat{\boldsymbol{n}}_1 \hat{\boldsymbol{b}}_2 \mathscr{I} + \hat{\boldsymbol{b}}_1 \hat{\boldsymbol{t}}_2 \mathscr{G}], \tag{6.63}$$

and for a source at Q_1

$$d\boldsymbol{H}(\mathrm{Q}_2) = Y_{\mathrm{c}}\, d\boldsymbol{p}_m \cdot [\hat{\boldsymbol{b}}_1 \hat{\boldsymbol{b}}_2 \mathscr{I}_{\mathrm{s}} + \hat{\boldsymbol{t}}_1 \hat{\boldsymbol{t}}_2 \mathscr{G}_{\mathrm{s}}], \tag{6.64}$$

where

$$\mathscr{I} = \mathrm{e}^{-jkt} \sqrt{\frac{d\eta_1}{d\eta_2}} \sum_{p=1}^{\infty} D_p^{\mathrm{h}}(1)\, A_p^{\mathrm{h}}(2)\, T_p^{\mathrm{h}} \tag{6.65a}$$

$$\mathscr{I}_{\mathrm{s}} = \frac{-jk\mathrm{e}^{-jkt}}{4\pi} \sqrt{\frac{1}{\varrho} \frac{d\psi_1}{d\psi_2}} \sum_{p=1}^{\infty} L_p^{\mathrm{h}}(1)\, A_p^{\mathrm{h}}(2)\, T_p^{\mathrm{h}} \tag{6.65b}$$

$$Y_{\mathrm{c}} = 1/Z_{\mathrm{c}},$$

and $\mathscr{G}$ and $\mathscr{G}_s$ have the same form as $\mathscr{I}$ and $\mathscr{I}_s$, respectively, except that the superscript h is replaced by s.

Employing reciprocity it is found that

$$A_p^{\mathrm{h}} = L_p^{\mathrm{h}} \tag{6.66a}$$

and

$$A_p^{\mathrm{s}} = -L_p^{\mathrm{s}}, \tag{6.66b}$$

which is not surprising when one recalls that the diffraction coefficient is the same for the excitation of a surface ray mode and the radiation from a surface ray mode.

If the surface source at Q_1 is an electric current dipole $\boldsymbol{P}_e = I l \hat{\boldsymbol{n}}_1$,

$$\boldsymbol{E}(P) = \hat{\boldsymbol{n}}_2 Z_c I l F_s \sqrt{\frac{\varrho}{s(\varrho + s)}}\, e^{-jks}, \tag{6.67}$$

with F_s given by (6.62), and the magnetic field induced on S is

$$\boldsymbol{H}(Q_2) = \hat{\boldsymbol{b}}_2 I l \mathscr{I}_s \tag{6.68}$$

with $\mathscr{I}_s$ given by (6.65b).

In the case of simple surfaces such as the spherical surface, the cylindrical surface, the conical surface and the plane surface, the geodesics are known and they are easy to describe; otherwise they can be found from the differential equations for geodesic paths, which is a formidable but straight forward exercise, see [6.26]. Calculating $d\eta_1$, $d\eta_2$, $d\psi_1$, $d\psi_2$ and ϱ is a matter of differential geometry involving the rays and the surface; this is discussed in [6.21, 26]. In the paragraphs to follow expressions will be given for the diffraction coefficients, launching coefficients and attenuation constants.

6.4.2. The Parameters

The diffraction coefficients, launching coefficients, and attenuation constants depend on the local geometry of the surface, the wave number k, and the nature of the surface, as described by the boundary conditions. KELLER and LEVY [6.21] have given the first-order terms in the expressions for the diffraction coefficients and attenuation constants. However, before we present their results, it is desirable to examine further the terms "soft" and "hard" boundary conditions.

This terminology is borrowed from acoustics. A soft boundary is one where the pressure field vanishes at the surface; it is also referred to as a Dirichlet boundary. On the other hand, a hard boundary is one where the normal derivative of the pressure field vanishes at the surface; this is also referred to as a Neumann boundary. Two types of surface ray modes have been assumed. For one type the electric field is in the binormal direction so that $\boldsymbol{E}_p = \hat{\boldsymbol{b}} E_p$, and for the other, the magnetic field is in the binormal direction so that $\boldsymbol{H}_p = \hat{\boldsymbol{b}} H_p$, and there is a normally-directed electric field $\hat{\boldsymbol{n}} \cdot \boldsymbol{E}_p$. The binormally directed electric field clearly satisfies a soft or Dirichlet boundary condition at the surface, whereas the binormally-directed magnetic field satisfies the boundary condition

$$\frac{\partial H}{\partial n} + \left(\frac{1}{h_b}\frac{\partial h_b}{\partial n}\right) H = 0, \tag{6.69}$$

in which h_b is the metrical coefficient associated with the unit vector $\hat{\boldsymbol{b}}$. The above boundary condition describes what we will refer to as a hard EM boundary. At high frequencies the second term is relatively small, so that the surface ray magnetic field satisfies a hard or Neumann boundary condition to a first approximation. Also (6.69) reduces to the hard boundary condition in the case of cylindrical surfaces where $h_b = 1$. These observations concerning the boundary conditions are of importance in the paragraphs to follow.

Let ϱ_g ne the radius of curvature of the surface along which the surface ray is propagating in the plane containing the normal to the surface and the tangent to the surface ray. As mentioned earlier, KELLER and LEVY [6.21] have used first-order asymptotic solutions for the diffraction of acoustic (scalar) and electromagnetic waves to deduce the attenuation constants and diffraction coefficients. For these canonical problems $\varrho_g = a$, a constant. According to KELLER and LEVY

$$\alpha_{0p}^{\mathrm{s}} = \frac{1}{a}\left(\frac{ka}{2}\right)^{1/3} q_p \,\mathrm{e}^{j\frac{\pi}{6}}, \tag{6.70}$$

$$[D_{0p}^{\mathrm{s}}]^2 = \frac{\mathrm{e}^{-j\frac{\pi}{12}}}{2^{5/6}\pi^{1/2}(ka)^{1/6}} \frac{a^{1/2}}{[\mathrm{Ai}'(-q_p)]^2} \tag{6.71}$$

for the soft surface, and

$$\alpha_{0p}^{\mathrm{h}} = \frac{1}{a}\left(\frac{ka}{2}\right)^{1/3} \bar{q}_p \,\mathrm{e}^{j\frac{\pi}{6}} \tag{6.72}$$

$$[D_{0p}^{\mathrm{h}}]^2 = \frac{\mathrm{e}^{-j\frac{\pi}{12}}}{2^{5/6}\pi^{1/2}(ka)^{1/6}} \frac{1}{\bar{q}_p} \frac{a^{1/2}}{[\mathrm{Ai}(-\bar{q}_p)]^2} \tag{6.73}$$

for the hard surface, where the Miller-type Airy function is given by

$$\mathrm{Ai}(-x) = \frac{1}{\pi}\int_0^{\infty} \cos(\tfrac{1}{3}t^3 - xt)\,dt\,,$$

$$\mathrm{Ai}(-q_p) = 0\,,$$

$$\mathrm{Ai}'(-\bar{q}_p) = 0\,,$$

and the prime denotes differentiation with respect to the argument of the function.

Table 6.2. Diffraction coefficients and attenuation constants for the curved surface

Surface	Square of diffraction coefficient $D_p^2 = (\text{Column A}) \cdot (\text{Column B})$	
	A. Keller's result	B. Correction terms
Soft acoustic and soft EM	$\dfrac{\pi^{-1/2}2^{-5/6}\varrho_g^{1/3}e^{-j\pi/12}}{k^{1/6}(\mathrm{Ai}'(-q_p))^2}$	$1+\left(\dfrac{2}{k\varrho_g}\right)^{2/3} q_p\left(\dfrac{1}{30}+\dfrac{\varrho_g}{4\varrho_{tn}}+\cdots\right)e^{-j\pi/3}$
Hard acoustic	$\dfrac{\pi^{-1/2}2^{-5/6}\varrho_g^{1/3}e^{-j\pi/12}}{k^{1/6}\boldsymbol{q}_p(\mathrm{Ai}(-\boldsymbol{q}_p))^2}$	$1+\left(\dfrac{2}{k\varrho_g}\right)^{2/3}\left(\boldsymbol{q}_p\left(\dfrac{1}{30}+\dfrac{\varrho_g}{4\varrho_{tn}}+\cdots\right) - \dfrac{1}{\boldsymbol{q}_p^2}\left(\dfrac{1}{10}+\dfrac{\varrho_g}{4\varrho_{tn}}+\cdots\right)\right)e^{-j\pi/3}$
Hard EM		$1+\left(\dfrac{2}{k\varrho_g}\right)^{2/3}\left(\boldsymbol{q}_p\left(\dfrac{1}{30}+\dfrac{\varrho_g}{4\varrho_{tn}}+\cdots\right) - \dfrac{1}{\boldsymbol{q}_p^2}\left(\dfrac{1}{10}-\dfrac{\varrho_g}{4\varrho_{tn}}+\cdots\right)\right)e^{-j\pi/3}$

ϱ_g = radius of curvature along the geodesic
ϱ_{tn} = radius of curvature perpendicular to the geodesic (transverse curve)
Dots indicate differentiation with respect to the arc length variable

VOLTMER [6.25] employing the same canonical problems as KELLER and LEVY, obtained attenuation constants and diffraction coefficients of improved accuracy by retaining higher-order terms in the asymptotic solutions. Voltmer's corrections to the attenuation constants and diffraction coefficients are of order $(2/ka)^{2/3}$.

The first-order approximations given by (6.70)–(6.73) do not depend on whether the surface is cylindrical or spherical or on whether the wave is acoustic or electromagnetic; however this is no longer the case with the more accurate formulas. An explanation of these second-order differences is best accomplished by examining the high-frequency diffraction from a more general surface, i.e., a surface of variable curvature along the ray path or of arbitrary curvature transverse to the ray path.

KELLER and LEVY [6.22], FRANZ and KLANTE [6.23], HONG [6.24], and VOLTMER [6.25] have considered the high-frequency diffraction by general convex surfaces. HONG has obtained asymptotic solutions to the integral equations for the plane wave diffraction by a hard acoustic surface and the plane wave diffraction by a hard EM boundary. VOLTMER has extended this work to soft boundaries, which are the same for acoustic and EM waves, as we have noted. The solutions were carried out to second order, and they are functions not only of ϱ_g, the radius of

Attenuation constant α_p = (Column C) · (Column D)		Zeroes of the Airy function
C. Keller's result	D. Correction terms	$\mathrm{Ai}(-q_p) = 0$ $q_1 = 2.33811$ $q_2 = 4.08795$
$\frac{q_p}{\varrho_g} e^{j\pi/6} \left(\frac{k\varrho_g}{2}\right)^{1/3}$	$1 + \left(\frac{2}{k\varrho_g}\right)^{2/3} q_p \left(\frac{1}{60} - \frac{2}{45} \varrho_g \ddot{\varrho}_g + \frac{4}{135} \dot{\varrho}_g^2\right) e^{-j\pi/3}$	$\mathrm{Ai}'(-q_1) = .70121$ $\mathrm{Ai}'(-q_2) = -.80311$
$\frac{\boldsymbol{q}_p}{\varrho_g} e^{j\pi/6} \left(\frac{k\varrho_g}{2}\right)^{1/3}$	$1 + \left(\frac{2}{k\varrho_g}\right)^{2/3} \left(\boldsymbol{q}_p \left(\frac{1}{60} - \frac{2}{45} \varrho_g \ddot{\varrho}_g + \frac{4}{135} \dot{\varrho}_g^2\right) + \frac{1}{\boldsymbol{q}_p^2} \left(\frac{1}{10} + \frac{\varrho_g}{4\varrho_{tn}} - \frac{\varrho_g \ddot{\varrho}_g}{60} + \frac{\dot{\varrho}_g^2}{90}\right)\right) e^{-j\pi/3}$	Zeroes of the derivative of the Airy function $\mathrm{Ai}'(-\boldsymbol{q}_p) = 0$
	$1 + \left(\frac{2}{k\varrho_g}\right)^{2/3} \left(\boldsymbol{q}_p \left(\frac{1}{60} - \frac{2}{45} \varrho_g \ddot{\varrho}_g + \frac{4}{135} \dot{\varrho}_g^2\right) + \frac{1}{\boldsymbol{q}_p^2} \left(\frac{1}{10} - \frac{\varrho_g}{4\varrho_{tn}} - \frac{\varrho_g \ddot{\varrho}_g}{60} + \frac{\dot{\varrho}_g^2}{90}\right)\right) e^{-j\pi/3}$	$\boldsymbol{q}_1 = 1.01879$ $\boldsymbol{q}_2 = 3.24820$ $\mathrm{Ai}(-\boldsymbol{q}_1 = .53566$ $\mathrm{Ai}(-\boldsymbol{q}_2) = -.41902$

curvature of the surface with respect to arc length along the ray trajectory, but also $\dot{\varrho}_g$, $\ddot{\varrho}_g$, and ϱ_{tn}, where the dot denotes a derivative with respect to arc length along the ray trajectory, and ϱ_{tn} is the radius of curvature of the surface in the direction of the binormal to the ray. Expressions for the attenuation constants are evident from the solutions; these are tabulated in Columns C and D of Table 6.2. On the other hand, complete expressions for the diffraction coefficients cannot be obtained from these solutions, because $\dot{\varrho}_g$ is assumed to be zero at the point of incidence on the surface, where the diffraction coefficient is evaluated. This condition was imposed to simplify the pertinent integral equations. The diffraction coefficients (more precisely, the diffraction coefficients squared) are given in Columns A and B of Table 6.2. The incomplete portion of the second-order term is indicated by (...); it is a function of $\dot{\varrho}_g$ and $\ddot{\varrho}_g$[4]. In deriving these results it is assumed that $\varrho_g/\varrho_{tn} < 1$: furthermore it is assumed that the surface rays have no torsion.

It is believed that the attenuation constants and diffraction coefficients listed in Table 6.2 are the best available at present and that they are adequate for most calculations, even though the expressions for the

[4] The $\dot{\varrho}_g$ and $\ddot{\varrho}_g$ terms in the diffraction coefficient will be the subject of a future investigation.

diffraction coefficients are not complete to second order. The improved attenuation constants are very important, because of the sensitivity of numerical calculations to errors in these parameters, particularly in the deep shadow region; corresponding errors in the diffraction coefficient are clearly less important to numerical accuracy.

The launching coefficients have been defined in (6.60). To determine the launching coefficients the radiation from a magnetic current moment on a perfectly-conducting sphere and the radiation from magnetic current line sources on cylindrical surfaces have been analyzed [6.27]. From the asymptotic solution of these canonical problems and their ray-optical interpretation it is found that for both cylindrical and spherical surfaces

$$L_p^{\mathrm{s}} = -\left(jk\frac{\pi}{2}\right)^{1/2} H_{\nu_p}^{(2)\prime}(ka)\, D_p^{\mathrm{s}}\,,$$

$$L_p^{\mathrm{h}} = -j\left(jk\frac{\pi}{2}\right)^{1/2} H_{\nu_p}^{(2)}(ka)\, D_p^{\mathrm{h}}\,,$$

where ν_p are the zeroes of the Hankel function in the first equation and the zeroes of the derivative of the Hankel function in the second equation. It is apparent that the relationship of the launching coefficient to the diffraction coefficient does not depend on the surface curvature transverse to the ray direction. For this reason, one may assume that

$$L_p^{\mathrm{s}} = \mathrm{e}^{-j\frac{\pi}{12}}(2\pi k)^{1/2}\left(\frac{2}{k\varrho_{\mathrm{g}}}\right)^{2/3} \cdot \mathrm{Ai}'(-q_p)\left[1-\left(\frac{2}{k\varrho_{\mathrm{g}}}\right)^{2/3}\frac{q_p}{15}\,\mathrm{e}^{j\frac{2\pi}{3}}\right]D_p^{\mathrm{s}}\,, \tag{6.74}$$

$$L_p^{\mathrm{h}} = \mathrm{e}^{j\frac{\pi}{12}}(2\pi k)^{1/2}\left(\frac{2}{k\varrho_{\mathrm{g}}}\right)^{1/3} \cdot \mathrm{Ai}(-q_p)\left[1+\left(\frac{2}{k\varrho_{\mathrm{g}}}\right)^{2/3}\frac{q_p}{15}\,\mathrm{e}^{j\frac{2\pi}{3}}\right]D_p^{\mathrm{h}}\,. \tag{6.75}$$

These expressions for the launching coefficients, where D_p^{s} and D_p^{h} are obtained from Table 6.2, are the best available at present. It would be desirable to solve a canonical problem where the magnetic current moment or line source is on a surface of variable curvature as a further check. The attachment coefficients follow from the above equations and (6.66).

From Table 6.2 it is seen that the real part of $\alpha_p^{\rm h} <$ the real part of $\alpha_p^{\rm s}$ so that $T_p^{\rm h}$ is exponentially larger than $T_p^{\rm s}$ and the F-type functions are exponentially larger than the G-type functions. Thus in the shadow region, the contributions from the latter functions are important only when $\boldsymbol{E}^i(1)$ is nearly parallel to $\hat{\boldsymbol{b}}_1$ or $d\boldsymbol{p}_{\rm m}$ is nearly parallel to $\hat{\boldsymbol{t}}_1$. Furthermore from examples involving a cylindrical geometry, it has been found that the dominant F-type functions are independent of torsion to a first-order approximation; however this is not the case for $G_{\rm s}$, $\mathcal{G}$, and $\mathcal{G}_{\rm s}$. Torsion appears to be a second-order effect, which is important mostly when accuracy is required in the deep shadow region.

6.4.3. Transition Regions

The series representations are rapidly convergent when the field point is deep in the shadow region. Usually only the first few terms are required to achieve reasonable accuracy, when the radii of curvature of the surface are larger than a wavelength or so. However as the field point Q_2 approaches close to Q_1 and more terms must be added to maintain accuracy, it is then no longer desirable to treat the excitation, propagation and diffraction of the different surface ray modes separately. As a result, the series representations are replaced by integral representations, and these are found to be proportional to Fock-type functions. When the source is on the surface the angular extent of the transition region from the shadow boundary is roughly $(k\varrho_{\rm g})^{-1/3}$; when the source is removed from the surface it is more nearly $(k\varrho_{\rm g}/2)^{-1/3}$.

In contrast with the deep shadow region a first-order asymptotic approximation is usually adequate for the transition region. In numerous calculations it was found that curves obtained from the expressions for the transition region joined smoothly with those obtained from expressions for the shadow region.

In describing the fields in the transition region, we employ the Fock-type Airy functions

$$w_1(\tau) = \frac{1}{\sqrt{\pi}} \int_{\Gamma_1} e^{\tau z - z^3/3}\, dz \tag{6.76}$$

and $w_2(\tau)$, which has the same form as $w_1(\tau)$ except that the contour of integration is Γ_2; these contours of integration are shown in Fig. 6.8. In the expressions to follow it will be convenient to use

$$\xi = \int_0^t \frac{1}{\varrho_{\rm g}} \left(\frac{k\varrho_{\rm g}}{2}\right)^{1/3} dt' . \tag{6.77}$$

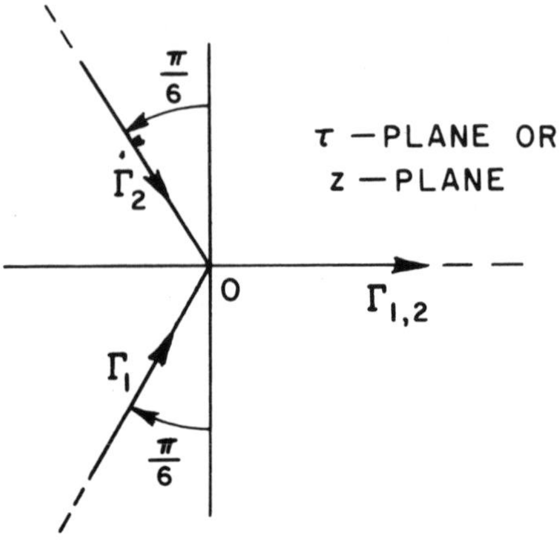

Fig. 6.8. Contours of integration for the Airy and Fock functions

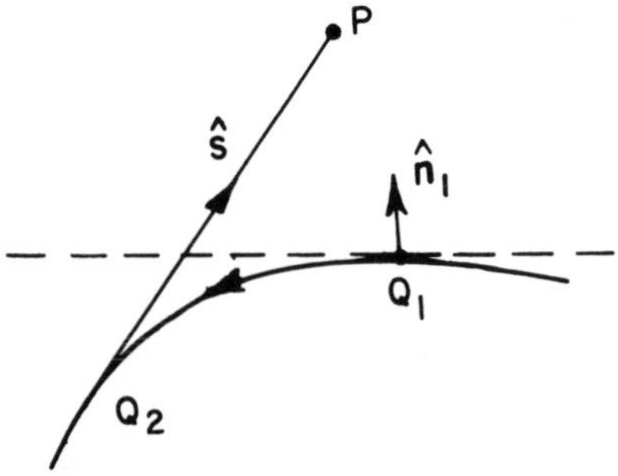

Fig. 6.9. A pseudo-ray system for calculating the field in the illuminated part of the transition region

In terms of ξ

$$T_p^{\mathrm{h}} = \mathrm{e}^{-j\xi\boldsymbol{\tau}_p} \quad \text{and} \quad T_p^{\mathrm{s}} = \mathrm{e}^{-j\xi\tau_p}$$

to first order, where $\boldsymbol{\tau}_p = \boldsymbol{q}_p \exp(-j\pi/3)$ and $\tau_p = q_p \exp(-j\pi/3)$.

When the field point is in the shadow region $\xi, t > 0$ and when the field point is in the illuminated region $\xi, t < 0$. In the illuminated portion of the transition region, one visualizes the surface ray as travelling from Q_1 to Q_2, where it sheds tangentially back toward P, as shown in Fig. 6.9. The ray path $\mathrm{Q}_1\mathrm{Q}_2\mathrm{P}$ does not obey the generalized Fermat's principle and therefore it is a pseudo-ray system, but it does serve as a useful coordinate system to calculate the field at P in the illuminated part of the transition region. Note that the surface ray divergence factors $\sqrt{d\eta_1/d\eta_2}$ and $\sqrt{d\psi_1/d\psi_2}$ are equal to one in this region.

As in the deep region, the expressions for the field in the transition region may be deduced from the solutions of the canonical problems for this region. On the other hand, recognizing the GTD solution for the deep shadow region as a residue series, it is sometimes possible to infer the integral representation from which it follows. However, this is a risky procedure and it appears to yield useful results only when the source is on the surface.

In the transition region

$$\sum_{p=1}^{\infty} L_p^{\mathrm{h}}(1)\, D_2^{\mathrm{h}}(2)\, T_p^{\mathrm{h}} \qquad \text{is replaced by}$$
$$g(\xi) \times \begin{cases} 1, & t \leqq 0 \\ x, & t \geqq 0 \end{cases}, \tag{6.78}$$

$$\sum_{p=1}^{\infty} L_p^{\mathrm{s}}(1)\, D_p^{\mathrm{s}}(2)\, T_p^{\mathrm{s}} \qquad \text{is replaced by}$$
$$-j\tilde{g}(\xi) \times \begin{cases} -\hat{\boldsymbol{n}}_1 \cdot \hat{\boldsymbol{s}}/\xi, & t \leqq 0 \\ x/m, & t \geqq 0 \end{cases}, \tag{6.79}$$

$$\sum_{p=1}^{\infty} L_p^{\mathrm{h}}(1)\, A_p^{\mathrm{h}}(2)\, T_p^{\mathrm{h}} \qquad \text{is replaced by}$$
$$\left(\frac{jk}{2}\right)^{1/2} (mx)^{-1}\, \psi(\xi), \qquad t > 0\,, \tag{6.80a}$$

$$\sum_{p=1}^{\infty} L_p^{\mathrm{s}}(1)\, A_p^{\mathrm{s}}(2)\, T_p^{\mathrm{s}} \qquad \text{is replaced by}$$
$$-\left(\frac{jk}{2}\right)^{1/2} (mx)^{-3}\, \tilde{\psi}(\xi), \qquad t > 0\,, \tag{6.80b}$$

where

$$\tilde{g}(\xi) = \frac{1}{\sqrt{\pi}} \int_{\Gamma_1} \frac{\mathrm{e}^{-j\xi\tau}}{w_2(\tau)}\, d\tau\,,$$

$$\psi(\xi) = \frac{1}{\sqrt{\pi}} \int_{\Gamma_1} \frac{w_2(\tau)}{w_2'(\tau)}\, \mathrm{e}^{-j\xi\tau}\, d\tau\,,$$

and

$$m = \left[\frac{k\varrho_g(1)}{2}\right]^{1/3},$$

$$x = \left[\frac{\varrho_g(2)}{\varrho_g(1)}\right]^{1/6}.$$

The function $g(\xi)$ has the same form as $\tilde{g}(\xi)$ except that $w_2(\tau)$ in the integrand is replaced by its derivative; moreover $\tilde{\psi}(\xi)$ follows from $\psi(\xi)$ when w_2'/w_2 is substituted for w_2/w_2'. The Fock-type functions above have been described and tabulated by LOGAN [6.28]; they are also briefly described in (Ref. [6.9], Section 1.3.3).

The expressions for the transition regions in (6.78) and (6.79) have been found to be very useful. Patterns calculated from them blend well with those calculated from geometrical optics deep in the illuminated region and from the surface ray modes in the shadow region [6.27]. On the other hand, (6.80a) and (6.80b) appear to be of limited value because the GTD approximation precludes Q_2 from being closer to Q_1 than a half wavelength or so. Moreover when the distance between Q_1 and Q_2 is small compared with ϱ_g, the currents or mutual coupling can be calculated with reasonable accuracy by assuming the surface is plane.

When the GTD solution involves a field reflected from a curved surface, the geometrical-optics representation of the reflected field is usually the greatest source of error. Furthermore the behavior of the field in the transition region adjacent to the shadow boundary is more complex than in the case of edge diffraction; this is particularly true in the illuminated part, where the solution should match the geometrical-optics approximation of the total field.

WAIT and CONDA [6.29] have obtained expressions for the field diffracted by a circular cylinder or sphere near the shadow boundary using a method based on the earlier work of FOCK [6.30] and GORIAINOV [6.31]. Their solution has been adapted (with slight modification) to the GTD format used here, with the result that for a surface whose ϱ_g is nearly constant in the transition region.

$$\sum_{p=1}^{\infty} D_p^{\mathrm{h}}(1)\, D_p^{\mathrm{h}}(2)\, T_p^{\mathrm{h}}$$

is replaced by (6.81)

$$D(x) - \left(\frac{2}{k}\right)^{1/2} m\, G(\xi)$$

with

$$D(x)=\sqrt{\frac{L}{\pi}}\,e^{j\left(\frac{\pi}{4}+x\right)}\,\mathrm{sgn}(\delta)\int\limits_{|\sqrt{x}|}^{\infty}e^{-j\tau^2}\,d\tau\,,$$

$$x=2kL\left(\frac{\delta}{2}\right)^2,$$

$$\xi=m\delta,$$

$$\delta=-\sin^{-1}\hat{\boldsymbol{n}}_1\cdot\hat{\boldsymbol{t}}_2\,,$$

$$L=\frac{ss'}{s+s'}\,,$$

for cylindrical and spherical waves normally incident on the shadow boundary. The quantities s' and s are the distances from the source and field points to their respective points of tangency on the curved surface, and for (6.81) to be valid $\sqrt{kL/2}$ should be greater than m.

The total field in the illuminated part of the transition region is the incident field of geometrical optics plus $\boldsymbol{E}^{\mathrm{d}}(P)$ given by (6.57), (6.58), and (6.81). $D(x)$ is the dominant term in (6.81); it is associated with the Fresnel (or Kirchhoff) diffraction by a half plane. As such it is independent of the polarization of the incident wave and the surface curvature. $G(\xi)$ can be regarded as a correction term containing this information; it has separate values for the s and h boundary conditions. The discontinuity in $D(x)$ at the shadow boundary $(x, \delta=0)$ exactly compensates the discontinuity in the incident field there so that the total high-frequency field is continuous at $\delta=0$. $G(\xi)$ is defined and presented graphically in [6.29]. Also $G(\xi)=\exp(-j\pi/4)\,[\tilde{p}(\xi)+1/(2\xi\sqrt{\pi})]$ and $\exp(-j\pi/4)\,[\tilde{q}(\xi)+1/(2\xi\sqrt{\pi})]$ for the s and h boundary conditions, respectively; the reflection coefficient functions $\tilde{p}$ and $\tilde{q}$ are described in [6.28] and (Ref. [6.9], Section 1.3.3).

This representation has been found to have good accuracy on the shadow boundary, and it can be shown that it blends with the GTD solution (6.57) for the deep shadow region. This blending it so first order; it is smoother for the h than the s boundary condition, and the larger kL the closer it occurs to the shadow boundary. Also this representation has been found to be quite accurate in the illuminated region near the shadow boundary, but it does not always join smoothly with the geometrical-optics field. Recently the author has become aware of the work of Ivanov [6.71] which appears to overcome this difficulty.

6.4.4. Two-Dimensional Problems

The preceding development can be applied to problems with a two-dimensional geometry by replacing

$$\sqrt{\frac{\varrho}{s(\varrho+s)}} \quad \text{with} \quad \frac{1}{\sqrt{s}},$$

$$\sqrt{\frac{d\eta_1}{d\eta_2}} \quad \text{or} \quad \sqrt{\frac{1}{\varrho}\,\frac{d\psi_1}{d\psi_2}} \quad \text{with} \quad 1,$$

$d\boldsymbol{p}_{\mathrm{m}}$ with $d\boldsymbol{M}$, a line of infinitesimal magnetic current, and $-jk/4\pi$ in (6.62) and (6.65b) with $-k\exp(j\pi/4)/\sqrt{8\pi k}$.

6.5. Applications

In applying the GTD one begins with ray tracing. The rays emanating from the primary source are considered first. The boundary surface of the radiating structure blocks the passage of the incident rays so that the space surrounding it is divided into an illuminated region occupied by the rays from the source and a shadow region where these rays do not penetrate. Upon encountering a perfectly-conducting surface the incident ray initiates a reflected ray from an interior point of the surface of a diffracted ray from edges, tips, or points of grazing incidence. The directions of these rays are determined by the laws of reflection and diffraction, which are corollaries to the generalized Fermat's principle; these laws have been described in the earlier sections. It should be noted that the reflected and diffracted rays also have illuminated regions which they cover and shadow regions which they do not penetrate. Moreover, these rays may encounter the boundary and give rise to higher-order diffracted or reflected rays. The field of the higher-order rays usually diminishes so rapidly with the number of successive diffractions that multiply-diffracted rays beyond the second or third order can be neglected. However in most two-dimensional problems (and some three-dimensional problems), it may be possible to sum the contributions from all the multiply-diffracted rays directly or to treat them by a self-consistent field procedure so that a closed form result is obtained.

Many problems are sufficiently simple so that the ray paths passing through a given field point can be determined without difficulty; in such cases the point of diffraction remains fixed as the field point varies or it moves about in a manner that can be described analytically. The paths of rays reflected from a plane surface are readily found by the method of images. However in the diffraction or reflection from a complex structure it may be necessary to employ a computer search routine to determine the points on the surface from which the rays emanate. These search procedures, such as the bisection method, employ the laws of diffraction or reflection.

It is evident that the GTD solution simplifies the surface radiation problem to the radiation from a finite array of scattering centers, which greatly reduces the time and cost of calculations. Since the GTD is a high-frequency method, these scattering centers cannot be spaced too closely. Generally speaking, they should be separated by a wavelength or more; however, the GTD solution often remains valid even for closer spacings. It is found that GTD solutions tend to fail gracefully as the frequency diminishes, until the low frequency or Rayleigh region is reached.

6.5.1. Reflector Antenna

The first example chosen to illustrate the application of the GTD is the calculation of the far-field pattern of an axially-symmetric parabolic reflector antenna. The essential components of our example are the reflector with a circular aperture of radius a and the feed positioned at its focus F, as shown in Fig. 6.10.

The geometrical optics far-field

$$\boldsymbol{E}^{\mathrm{go}} = \begin{cases} \boldsymbol{E}^{\mathrm{f}}, & \pi - \alpha > \theta > \delta \\ 0, & \pi - \alpha < \theta \leqq \pi, \end{cases} \tag{6.82}$$

where $\boldsymbol{E}^{\mathrm{f}}$ is the field of the feed, which is either calculated or measured and θ is the polar angle measured from the z-axis. A small angular sector $\delta \approx 5(ka)^{-1}$ in the forward axial direction has been excluded from the GTD solution, because of the difficulties encountered when there is a confluence of a reflection boundary and a caustic of the diffracted rays. The field in this region can be determined by the current-distribution method or the aperture-field method (Ref. [6.14], Chapters 6 and 9) and [6.32].

Let us calculate the pattern at the point P in the plane ϕ, $\pi + \phi$. This plane intersects the edge of the reflector at 1 and 2. Diffracted rays are induced at each point on the edge of the reflector, but only the straight line paths joining F and P which pass through the Points 1 and 2 satisfy (6.28). The paths F1P and F2P are the minimum and maximum distances between F and P which include a point on the edge of the reflector.

The ray singly-diffracted from edge 2 crosses the aperture and induces a doubly-diffracted ray at edge 1. The contribution to the far field from

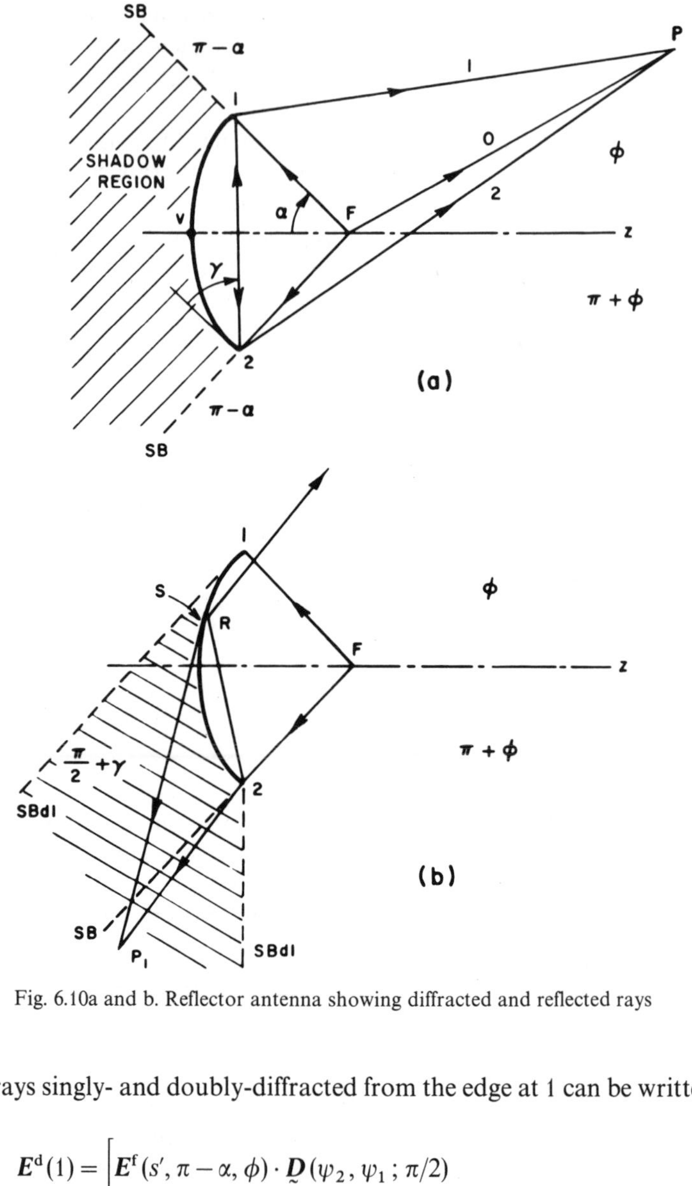

Fig. 6.10a and b. Reflector antenna showing diffracted and reflected rays

the rays singly- and doubly-diffracted from the edge at 1 can be written as

$$\begin{aligned} \boldsymbol{E}^{\mathrm{d}}(1) = \Big[& \boldsymbol{E}^{\mathrm{f}}(s', \pi-\alpha, \phi) \cdot \underset{\sim}{\boldsymbol{D}}(\psi_2, \psi_1; \pi/2) \\ & + \boldsymbol{E}^{\mathrm{f}}(s', \pi-\alpha, \phi+\pi) \cdot \underset{\sim}{\boldsymbol{D}}(\gamma, \psi_1; \pi/2) \cdot \underset{\sim}{\boldsymbol{D}}(\psi_2, \gamma; \pi/2) \\ & \cdot j \frac{\mathrm{e}^{-j2ka}}{\sqrt{2a}} \Big] \sqrt{\frac{a}{\sin\theta}} \, \frac{\mathrm{e}^{-j(kR - ka\sin\theta)}}{R} \end{aligned} \qquad (6.83)$$

in which

$$\psi_1 = \gamma + \frac{\pi}{2} - \alpha\,,$$

$$\psi_2 = \gamma + \frac{\pi}{2} + \theta\,,$$

s' is the distance from F to the edge of the reflector, and R is the distance from P to the center of the aperture, which is chosen as the phase reference.

The factor j appears in the second term because there is a caustic on the ray which crosses the aperture. The ray diffracted from 1 is shadowed in the region $\pi/2 < \theta < \pi/2 + \gamma$, $\phi + \pi$ shown in Fig. 6.10b. The discontinuity in the geometrical-optics field at the shadow boundary $(\pi - \alpha, \phi)$ is compensated by the discontinuity in the field of the ray singly-diffracted from 1, and the discontinuity in the field of the ray singly-diffracted from 2 is compensated by the field of the ray doubly-diffracted from 1. The field $\boldsymbol{E}^{\mathrm{d}}(2)$ of the rays singly- and doubly-diffracted from 2 has a similar form. The fields of the higher-order multiply-diffracted rays could be included in the solution, but this contribution is insignificant when the aperture diameter is greater than a few wavelengths.

The field of the ray diffracted from the edge at 2 and then reflected from R on the concave side of the reflector is given by

$$\boldsymbol{E}^{\mathrm{dr}}(2, R) = -\boldsymbol{E}^{\mathrm{f}}(s', \pi - \alpha, \phi + \pi) \cdot \underset{\sim}{\boldsymbol{D}}(\psi_r, \psi_1\,; \pi/2) \cdot \underset{\sim}{\boldsymbol{R}} \sqrt{\frac{|\varrho|}{s(\varrho + s)}}\,\sqrt{|\varrho_1 \varrho_2|}\, e^{-jks}\, \frac{\mathrm{e}^{-j(kR + \phi_{\mathrm{R}})}}{R}\,, \tag{6.84}$$

$$\theta_2 < \theta < \theta_1$$

in which ψ_{r} is the angle between the ray diffracted to R and the plane tangent to the reflector at 2, s is the distance between 2 and R, ϱ is the caustic distance of the edge diffracted ray (it is negative), ϱ_1, ϱ_2 are the principal radii of curvature of the reflected wavefront at R, f is the focal distance of the reflector, ϕ_{R} is the phase factor relating this contribution to the phase center, $\theta_1 = 2\tan^{-1}(2f/a) - \pi/2$, and $\theta_2 = \tan^{-1}(4f/a)$.

A caustic occurs between 2 and R and a second caustic between R and P, which accounts for the minus sign preceding $\boldsymbol{E}^{\mathrm{f}}$. $\boldsymbol{E}^{\mathrm{dr}}(2, \mathrm{R})$ vanishes outside the interval $\theta_2 < \theta < \theta_1$. θ may be expressed in terms of ψ_{r} but in calculating the pattern, we wish to determine ψ_{r} (and the Point R) as a function of θ. This can be done by a simple computer search routine such as the bisection method mentioned earlier. Equations (6.83) and

(6.84) simplify markedly in the H- and E-plane, where $\boldsymbol{E}^{\mathrm{f}}$ is usually parallel or perpendicular to the edge at 1 and 2. For example, in the H-plane $\boldsymbol{E}^{\mathrm{f}} \cdot \underset{\sim}{\boldsymbol{D}} = -\boldsymbol{E}^{\mathrm{f}} \cdot \underset{\sim}{\boldsymbol{D}} \cdot \underset{\sim}{\boldsymbol{R}} = E^{\mathrm{f}} D_{\mathrm{s}}$ and in the E-plane $\boldsymbol{E}^{\mathrm{f}} \cdot \underset{\sim}{\boldsymbol{D}} = \boldsymbol{E}^{\mathrm{f}} \cdot \underset{\sim}{\boldsymbol{D}} \cdot \underset{\sim}{\boldsymbol{R}} = -E^{\mathrm{f}} D_{\mathrm{h}}$. There is a similar ray diffracted from edge 1 and reflected from the concave surface.

In the region $\pi - \delta > \theta > \delta$

$$\boldsymbol{E}(P) = \boldsymbol{E}^{\mathrm{go}} + \boldsymbol{E}^{\mathrm{d}}(1) + \boldsymbol{E}^{\mathrm{d}}(2) + \boldsymbol{E}^{\mathrm{dr}}(1, \mathrm{R}) + \boldsymbol{E}^{\mathrm{dr}}(2, \mathrm{R}) . \tag{6.85}$$

A more complete analysis would include the surface rays exctited at 1 and 2; however the contribution from these rays is generally very weak.

The rear axis of the reflector is a caustic of the diffracted rays, so the GTD can not be used here without modification. In the caustic region $\pi - \delta < \theta \leqq \pi$, the field can be calculated from an integral representation using the equivalent electric

$$I(\phi) = \frac{-\hat{e} \cdot \boldsymbol{E}^{\mathrm{f}}}{Z_{\mathrm{c}}} D_{\mathrm{s}}(2\pi - \gamma, \psi_1 ; \pi/2) \sqrt{\frac{8\pi}{k}}\, \mathrm{e}^{-j\frac{\pi}{4}} \tag{6.86}$$

and magnetic

$$M(\phi) = -(\hat{e} \times \hat{s}') \cdot \boldsymbol{E}^{\mathrm{f}} D_{\mathrm{h}}(2\pi - \gamma, \psi_1 ; \pi/2) \sqrt{\frac{8\pi}{k}}\, \mathrm{e}^{-j\frac{\pi}{4}} \tag{6.87}$$

ring currents flowing on the dege of the reflector [6.32–6.34]. This procedure is accurate for $\theta = \pi$, and it is a good approximation in the caustic region joining smoothly with the field of the two edge diffracted rays, if the diffraction coefficient is slowly varying in this region.

AFIFI [6.35] has measured the H-plane pattern of a parabolic reflector antenna mounted on a ground plane and fed by a monopole. This is a desirable configuration to test our solution, because the scattering from the feed support has been neglected. The measured pattern and the pattern calculated from our GTD solution are shown in Fig. 6.11. The two patterns are seen to be in good agreement. In the range of aspects $10^0 < \theta < 70^0$ the pattern is quite frequency sensitive. If the frequency used to calculate the pattern is changed by only 5 percent, the agreement between the calculated and measured patterns is greatly improved.

The wide angle side lobes also can be calculated by physical optics, but the computational time is much greater and the results are less accurate. The GTD can be applied to calculate the patterns of a wide class of reflector antennas including those with subreflectors, where the scattering from the subreflector must be taken into account.

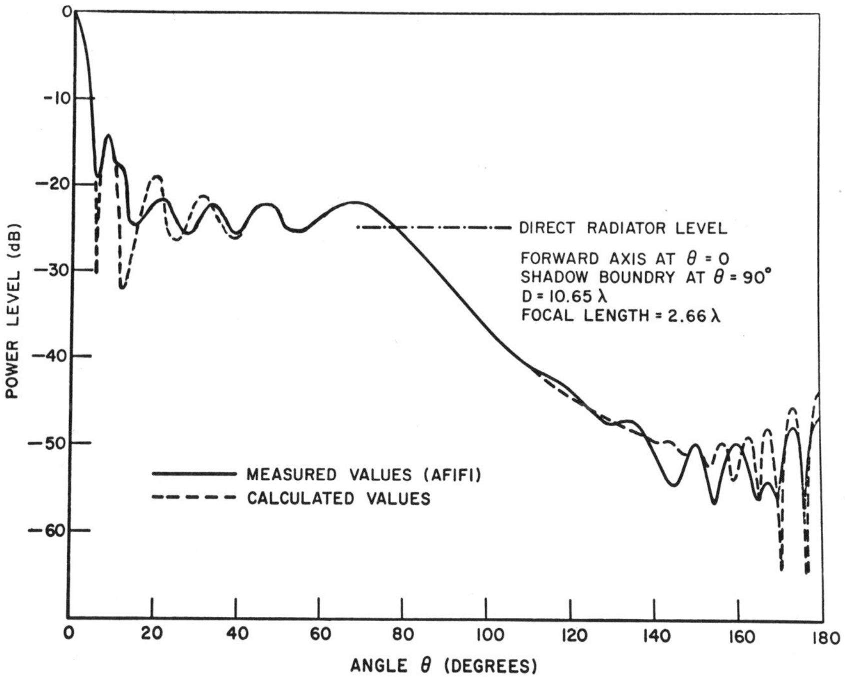

Fig. 6.11. H-plane pattern of a parabolic reflector antenna with a dipole feed

6.5.2. Slot in an Elliptic Cylinder

As a second example of the application of the GTD we will briefly consider the radiation from an axial slot in a perfectly-conducting elliptic cylinder. The far-zone pattern is to be calculated in a plane perpendicular to the axis of the slot so that the surface rays involved are torsionless. The rays which contribute to the field in the illuminated and shadow regions are depicted in Fig. 6.12. The equation

$$y_\mathrm{d} = \pm a \Bigg/ \sqrt{1 + \left(\frac{b}{a}\right)^2 \tan^2 \phi} \tag{6.88}$$

locates the points of diffraction Q_2 and Q_2' as a function of the aspect angle ϕ, which together with b and a is defined in Fig. 6.12. Equations (6.61), (6.62), and (6.78) and the expressions for the GTD parameters and $g(\xi)$ are used to calculate the diffracted field. The finite width of the slot (0.34 wavelength) is taken into account by an array of 5 magnetic dipoles in the aperture. In Fig. 6.13 calculated patterns are compared with

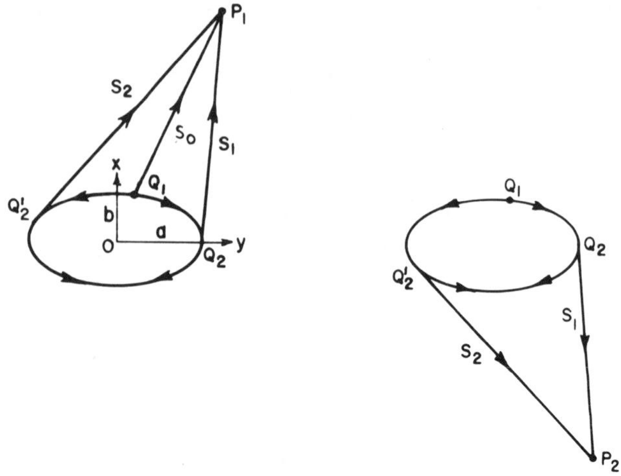

Fig. 6.12. Rays emanating from a source on a curved surface

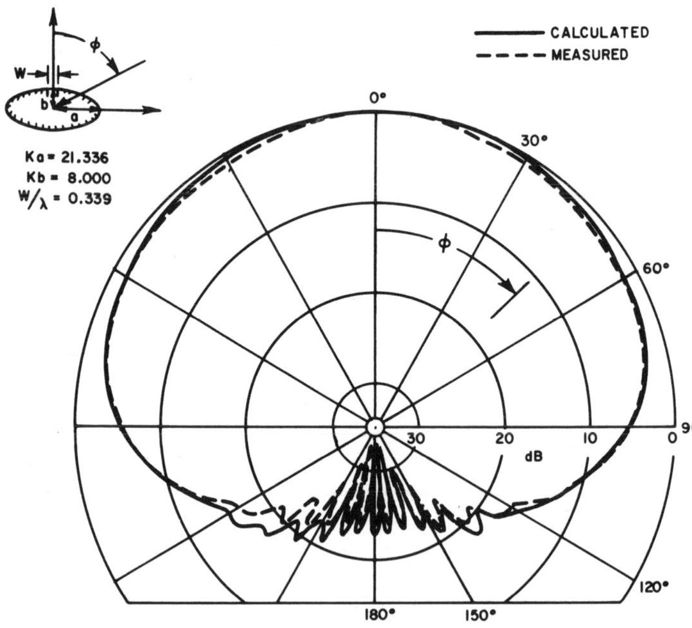

Fig. 6.13. Pattern of an axial slot on an elliptic cyclinder

measured patterns [6.27], and the agreement between the two patterns is seen to be good. The discrepancy between the calculated and measured patterns is no worse than that between the two halves of the measured pattern about its vertical axis.

As noted earlier, the formulas given in this chapter can be applied formally to cases where there are torsional surface rays, such as a pattern calculation in an oblique plane in the present example. Since the dominant term in (6.61) containing F_s is independent of torsion to first order, the torsional effects are not too noticeable except at the lower levels of the patterns. On the other hand, in calculating the mutual coupling between slots, which is exclusively a shadow region phenomenon, the torsional effects are much more evident than in the corresponding radiation pattern.

In the two examples considered thus far the GTD has been used to calculate far fields, but there is no reason why it cannot be used to calculate near fields provided that the field point is not too close to a point of diffraction, which is roughly a wavelength in the case of edge diffraction.

6.5.3. Monopole Antenna Near an Edge

Consider a monopole radiating in the presence of an edge, as shown in Fig. 6.14. This example is chosen to demonstrate the flexibility of the GTD by combining it with the moment method described in the earlier chapters. For simplicity pulse basis functions and impulse test functions are employed. Dividing the monopole into N segments of length Δ, the moment method solution is given compactly by

$$-E^{\mathrm{i}}_m = \sum_{n=1}^{N} Z_{mn} I_n, \qquad m = 1, 2, \ldots, N, \tag{6.89}$$

where E^{i}_m is the axial component of the incident electric field at the center of the mth segment, I_n is the constant current on the nth segment, and Z_{mn} is an element of the generalized impedance matrix, which describes the total interaction between the m and nth segments.

Let the current on the nth segment be the source of the electric field $\boldsymbol{E}_m$ at the center of the mth segment. The impedance element $Z_{mn} = E_{zm}/I_n$ in which E_{zm} is the axial (or z) component of $\boldsymbol{E}_m$.

$$E_{zm} = E^{\mathrm{o}}_{zm} + E^{\mathrm{r}}_{zm} + E^{\mathrm{d}}_{zm}, \tag{6.90}$$

where E^{o}_{zm} is due to the free-space radiation from I_n and E^{r}_{zm} is due to reflection from the horizontal surface of the wedge, which may be found from the image of I_n as indicated in Fig. 6.14. It follows then that for

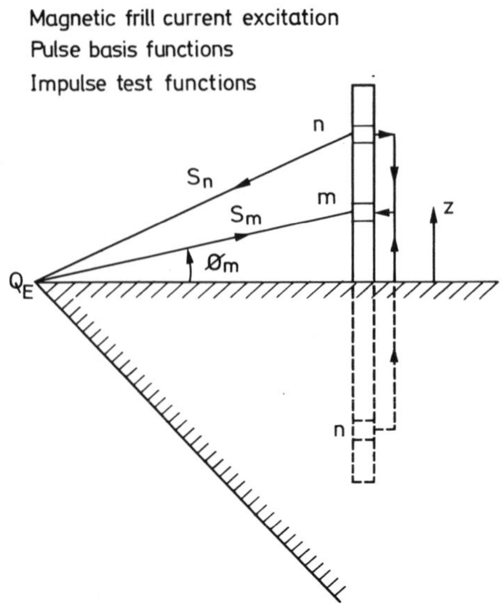

Fig. 6.14. Monopole on a wedge

$\Delta \ll \lambda$ and $|z_n - z_m|$,

$$Z^{\mathrm{o}}_{mn} + Z^{\mathrm{r}}_{mn} \approx \frac{jkZ_{\mathrm{c}}\Delta}{4\pi}\left[1 - \frac{1}{k^2}\frac{\partial^2}{\partial z\,\partial z'}\right]\frac{\mathrm{e}^{-jkr}}{r}\Bigg|_{\substack{z=z_m\\ z'=\pm z_n}} \tag{6.91}$$

with $r = \sqrt{(z-z')^2 + a^2}$ in which a is the radius of the monopole; $z' = \pm z_n$ implies that (6.91) has two terms.

The axial component of the diffracted electric field E^{d}_{sm} is determined by the GTD. The pertinent ray path is $n\mathrm{Q_E}m$, and

$$Z^{\mathrm{d}}_{mn} = \frac{jkZ_{\mathrm{c}}\Delta}{4\pi}\,\frac{D_{\mathrm{h}}\left(\phi_m, \phi_n; \dfrac{\pi}{2}\right)}{\sqrt{S_n S_m (S_n + S_m)}}\,\mathrm{e}^{-jk(S_n+S_m)}\cos\phi_m \cos\phi_n\,. \tag{6.92}$$

The current at N points on the monopole is given by the current column matrix

$$[I] = [Z]^{-1}\,[E^{\mathrm{i}}]$$

in which $[Z]^{-1}$ is the inverse of the generalized impedance matrix and $[E^{\mathrm{i}}]$ is the column matrix with N elements E^{i}_m. The GTD-moment

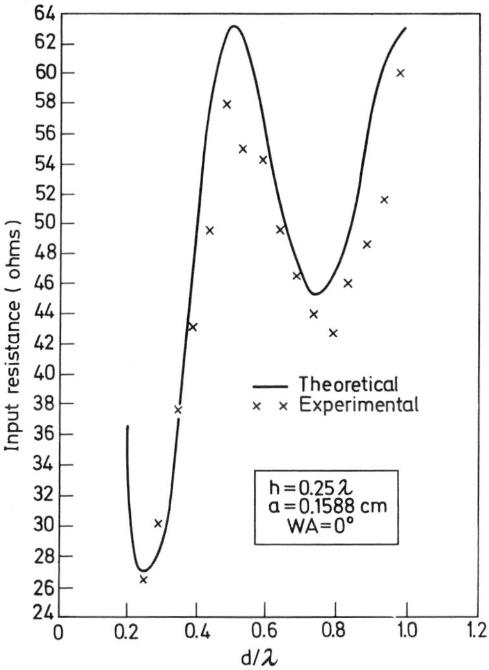

Fig. 6.15. Input resistance of a monopole at the center of an octagonal plate

method solution described here is based on the work of THIELE and NEWHOUSE [6.36], who applied it to calculate the input impedance to a monopole at the center of square, octagonal and circular plates. The coaxial feed is modelled by a magnetic frill current (Ref. [6.37], Appendix I), which is the source of E^i_m, and in these examples the wedge angle $\mathrm{WA} = (2-n)\,\pi = 0$. In Fig. 6.15 calculated values of the input resistance for the octagonal plate case are compared with measured values. The agreement is reasonably good considering the difficulties encountered with the measurements. The measured and calculated values of the input reactance compare similarly. The length of the monopole is h, its radius is a, and the perpendicular distance from its base to the sides of the octagonal plate is d.

It is clear that this hybrid GTD-moment method can be applied to a general wire antenna configuration. Moreover it can be readily extended to a receiving wire antenna. In this case, referring to the geometry in Fig. 6.14, E^i_m consists of directly-radiated, reflected and edge-diffracted terms.

6.5.4. Discussion

The examples presented in the preceding subsections are only a small sample of the problems that have been treated by the GTD. In the early history of the theory KELLER and his co-workers applied the GTD to determine the scalar diffraction by slits, circular apertures, cylinders, cylinder tipped-half planes, cones, spheres, and spheroids; in addition they treated the electromagnetic diffraction by spheres, cones, and arbitrary cylinders illuminated at oblique incidence. In their papers the pertinent ray analysis is given along with expressions for the diffracted fields away from the radiating body; however the solutions are not valid in the transition regions adjacent to shadow and reflection boundaries. A convenient review of these papers is given in [6.12]; also many of their results are described in [6.9] along with some additional results for the diffraction by a strip. An improved GTD solution for the diffraction by a cylinder-tipped half plane using the diffraction coefficients and attenuation constants in Table 6.2 is described in [6.38]. Additional contributions to the electromagnetic backscatter from cones are presented in [6.33, 6.39–6.43], but the GTD analysis is still not complete. For example, work remains to be done on the base-tip interaction, particularly with regard to the transition phenomena associated with the shadowing of the base by the tip. The electromagnetic backscatter by circular discs is treated in [6.44, 6.45]. The solution reported in the latter reference contains a type of doubly-diffracted ray omitted in the former; however the treatment of the fields of the doubly-diffracted rays on the illuminated and shadow side of the disc and the factor of 1/2 required at grazing incidence is obscure. The backscatter from rectangular plates is described in [6.46], where the problem is reduced to that of a strip for the GTD analysis. In [6.47] a GTD solution is presented for the scattering by a rectangular cylinder which is illuminated by the field of a line source. Since the edge diffraction coefficients in (6.33) are used, the solution is valid in the transition regions. Calculated patterns are found to be in excellent agreement with those calculated by the moment method. A comparison of GTD and moment method solutions for the strip and the two-dimensional corner and trough is carried out in [6.48]; the patterns calculated from the GTD solutions, which are valid in the transition regions, are in excellent agreement with those obtained by the moment method, for the intermediate-sized structures (0.5 to 5 λ). The GTD has also been applied to calculate the scattering from thin, curved wires [6.49]. The patterns are found to be in good agreement with those calculated by the variational method.

The GTD has been applied to a variety of antenna problems. The parabolic reflector antenna is treated in [6.32], and calculations of the scattering from a hyperboloidal subreflector are presented in Chapter 7; Subsection 7.3.4. Ray-optical methods are employed to determine the aperture reflection coefficients and radiation patterns of parallel-plate waveguides in [6.50–6.55]. The related problem of the E- and H-plane patterns of horn antennas was studied in [6.56–6.58]. The calculation of the patterns in the principal planes is facilitated by reducing the geometry to a two-dimensional configuration; however in the H-plane calculation, one may also get significant contributions from rays diffracted from the edges parallel to this plane. The reflection coefficient for the junction of a rectangular waveguide and an H-plane sectorial horn is calculated in [6.59]. The diffraction coefficient in (6.33) is needed to get accurate results for very small flare angles where overlapping transition regions occur at the junction. In [6.60] the GTD is used to investigate the gain and radiation pattern of a conical horn excited by a circular waveguide operating in the TE_{11} mode.

The examples discussed thus far have involved relatively simple shapes, but the GTD also can be used to calculate the radiation from complex structures. The radiation pattern of a linerar array of pistons mounted in one face of a rigid, rectangular box is considered in [6.61]. Although this is an acoustics problem, it has much in common with an array of slots in a rectangular ground plane. When the contributions from the incident (geometrical optics) ray plus all the rays singly- and doubly-diffracted from the 12 edges of the box are taken into account, the resulting calculated patterns are in excellent agreement with measured patterns. The GTD has been employed to calculate the patterns of slots and monopoles positioned on the fuselage of an aircraft [6.26, 6.62]. Even though the aircraft is modelled in its most basic form, so that only the fuselage and wings are considered in the roll plane analysis, the ray analysis is not simple. In addition to the incident ray, a surface ray is launched on the fuselage, and the surface ray sheds a diffracted ray, which in turn may be reflected and diffracted from the wings. Generally speaking, the agreement between calculated and measured patterns is excellent. The GTD has been used to predict the patterns of satellite antennas [6.63, 6.64]. The configurations studied consist of dipoles radiating in the presence of finite circular and rectangular cylinders and monopoles mounted on the cylinder. Calculated patterns, largely based on preliminary results, appear to be in quite good agreement with measured patterns. The edge diffraction coefficients described in this chapter are employed in [6.26, 6.61–6.64].

For simplicity the discussion in this chapter has been restricted to the radiation from hard, soft and perfectly-conducting surfaces in isotropic, homogeneous media, but the GTD can be extended to treat bodies in inhomogeneous media [6.65, 6.66] and anisotropic media

[6.67] and to bodies with penetrable [6.68] and impedance surfaces (Ref. [6.9], pp. 48–49, and [6.28]). A thorough treatment of ray-optical fields in inhomogeneous and anisotropic media is given in [6.3, 6.69].

6.6. Conclusions

In the geometrical theory of diffraction, the diffracted fields propagate in the same manner as the geometrical-optics field. The spreading of rays in a plane containing the edge caustic or the caustic or surface diffraction is determined from differential geometry; thus the field of a diffracted ray is determined in part by geometrical considerations and in part by wave considerations provided by the diffraction coefficients. Furthermore, it is assumed that the diffraction mechanism is, in effect, localized at an edge or shadow boundary, and that it functions independently of the other parts of the structure. The ultimate accuracy of the geometrical theory of diffraction appears to be limited by these two assumptions, particularly in the case of the smaller radiating structures.

Numerous comparisons of GTD solutions with asymptotic expansions, calculations based on convergent (exact) methods, and measurements have shown that the GTD provides a systematic approach to the high-frequency solution of a wide variety of antenna, scattering and propagation problems. In many instances the GTD solution is not only accurate at high frequencies, but also at relatively low frequencies, where the ratio of the characteristic dimension to wavelength is of order unity. However, presuming that the GTD is an asymptotic method, one expects it to fail as the frequency is reduced, regardless of the number of terms retained in the approximation. At sufficiently low frequencies the local behavior of reflection and diffraction break down.

As a purely ray-optical technique the GTD fails near caustics and in transition regions adjacent to shadow and reflection boundaries; however in this chapter we have shown that the GTD can be extended to calculate fields in the transition regions. As noted earlier, supplementary methods exist for treating the field at a caustic; often one can introduce an integral representation of the field where geometrical optics or the GTD is used to determine the equivalent source. If desired, the accuracy of the geometrical optics current can be improved by using Ufimstev's method [6.70].

The reason for using the GTD method stems from the significant advantages to be gained; namely

a) it is simple to use, and yields accurate results;

b) it provides some physical insight into the radiation and scattering mechanisms involved;

c) it can be used to treat problems for which exact analytical solutions are not available;

d) it can be combined readily with other methods such as the moment method.

References

6.1. R.K. Luneberg: *Mathematical Theory of Optics* (Brown University Notes Providence, 1944). Also published by (University of California Press, Berkeley, 1964).
6.2. M. Kline: In *The Theory of Electromagnetic Waves* (Interscience Publishers Inc., New York, 1951), pp. 225–262. See also his chapter in *Electromagnetic Waves*, ed. by R.E. Langer (University of Wisconsin Press, Madison, 1962), pp. 3–32.
6.3. M. Kline, I. Kay: *Electromagnetic Theory and Geometrical Optics* (Interscience Publishers, New York, 1965).
6.4. J.B. Keller, R.M. Lewis, B.D. Seckler: Comm. Pure Appl. Math. **9**, 208 (1973).
6.5. R.G. Kouyoumjian: Proc. IEEE **53**, 864 (1965).
6.6. I. Kay, J.B. Keller: J. Appl. Phys. **25**, 876 (1954).
6.7. D. Ludwig: Comm. Pure Appl. Math. **19**, 215 (1966).
6.8. R.G. Kouyoumjian, P.H. Pathak: Proc. IEEE **62**, 1448 (1974).
6.9. J.J. Bowman, T.B.A. Senior, P.L.E. Uslenghi: *Electromagnetic and Acoustic Scattering by Simple Shapes* (North-Holland Publishing Co., Amsterdam, 1969).
6.10. J.B. Keller: In "Proc. Symposium on Microwave Optics", p. II, McGill University (1953) Astia Document AD 211500 (1959), pp. 207–210.
6.11. J.B. Keller: In *Calculus of Variations and its Applications*, ed. by L.M. Graves (McGraw-Hill Book Co., New York, 1958), pp. 27–52.
6.12. J.B. Keller: J. Opt. Soc. Am. **52**, 116 (1962).
6.13. M. Born, E. Wolf: *Principles of Optics*, 4th ed. (Pergamon Press, Oxford, 1970), pp. 753, 754.
6.14. S. Silver: *Microwave Antenna Theory and Design* (McGraw Hill Book Co., New York, 1949), pp. 119–122.
6.15. P.H. Pathak, R.G. Kouyoumjian: "The Dyadic Diffraction Coefficient for a Perfectly-Conducting Wedge", Report 2183-4, Electro-Science Laboratory, The Ohio State University, Columbus, Ohio 1973).
6.16. S.N. Karp, J.B. Keller: Optica Acta **8**, 61 (1961).
6.17. L. Kaminetzky, J.B. Keller: SIAM J. Appl. Math. **22**, 109 (1972).
6.18. T.B.A. Senior: IEEE Trans. Antennas Propagation AP-**20**, 326 (1972).
6.19. K.M. Siegel, J.W. Cripsin, E. Schensted: J. Appl. Phys. **26**, 309 (1955).
6.20. L.B. Felsen: IRE Trans. Antennas Propagation AP-**5**, 121 (1957).
6.21. B.R. Levy, J.B. Keller: Comm. Pure Appl. Math. **12**, 159 (1959).
6.22. J.B. Keller, B.R. Levy: IRE Trans. Antennas Propagation AP-**7**, S52.
6.23. W. Franz, K. Klante: IRE Trans. Antennas Propagation AP-**7**, S68 (1959).
6.24. S. Hong: J. Math. Phys. **8**, 1223 (1967).
6.25. D.R. Voltmer: "Diffraction by Doubly Curved Convex Surfaces", Ph.D. Dissertation, The Ohio State Univ., Columbus, Ohio (1970).
6.26. W.D. Burnside: "Analysis of On-Aircraft Antenna Patterns", Report 3390-1, Electro-Science Laboratory, The Ohio State Univ., Columbus, Ohio (1972).
6.27. P.H. Pathak, R.G. Kouyoumjian: Proc. IEEE **62**, 1438 (1974).
6.28. N.A. Logan: "General Research in Diffraction Theory", Vols. I, II, Reports LMSD-288087, 288088, Lockheed Aircraft Corporation, Missiles and Space Division, Sunnyvale, Calif. (1959). See also N.A. Logan, K.S. Yee: In *Electro-*

magnetic Waves, ed. by R.E. LANGER (University of Wisconsin Press, Madison, 1962), pp. 139–180.
6.29. J.R. WAIT, A.M. CONDA: J. Res. of N.B.S. **63**D, 181 (1959).
6.30. V.A. FOCK: *Electromagnetic Diffraction and Propagation Problems* (Pergamon Press, Oxford, 1965), pp. 134–146.
6.31. A.S. GORIAINOV: Radio Eng. Electron. (USSR) **3**, 23 (1958) English translation of Radiotechn. i Elektron. **3**, 603 (1958).
6.32. P.A.J. RATNASIRI, R.G. KOUYOUMJIAN, P.H. PATHAK: "The Wide Angle Side Lobes of Reflector Antennas", Report 2183-1, Electro-Science Laboratory, The Ohio State Univ., Columbus, Ohio (1970).
6.33. C.E. RYAN, JR., L. PETERS, JR.: IEEE Trans. Antennas Propagation AP-**17**, 292 (1969).
6.34. C.E. RYAN, JR., L. PETERS, JR.: IEEE Trans. Antennas Propagation AP-**18**, 275 (1970).
6.35. M.S. AFIFI: In *Electromagnetic Wave Theory*, Part 2, ed by J. BROWN (Pergamon Press, Oxford, 1967), pp. 669–687.
6.36. G.A. THIELE, T.H. NEWHOUSE: IEEE Trans. Antennas Propagation AP-**23**, 62 (1975).
6.37. G.A. THIELE: In *Computer Techniques in Electromagnetics*, ed. by R. MITTRA (Pergamon Press, Oxford, 1973), pp. 7–95.
6.38. R.G. KOUYOUMJIAN, W.D. BURNSIDE: IEEE Trans. Antennas Propagation AP-**18**, 424 (1970).
6.39. M.E. BECHTEL: Proc. IEEE **53**, 877 (1965).
6.40. R.A. ROSS: IEEE Trans. Antennas Propagation AP-**17**, 241 (1969).
6.41. T.B.A. SENIOR, P.L.E. USLENGHI: Radio Science **6**, 393 (1971).
6.42. W.D. BURNSIDE, L. PETERS, JR.: Radio Science **7**, 943 (1972).
6.43. T.B.A. SENIOR, P.L.E. USLENGHI: Radio Science **8**, 247 (1973).
6.44. R.V. DEVORE, R.G. KOUYOUMJIAN: "The Back Scattering from a Circular Disk", URSI-IRE Spring Meeting, Washington, D.C. (1961).
6.45. E.F. KNOTT, T.B.A. SENIOR, P.L.E. USLENGHI: Proc. IEE (London) **118**, 1736 (1971).
6.46. R.A. ROSS: IEEE Trans. Antennas Propagation AP-**14**, 329 (1966).
6.47. R.G. KOUYOUMJIAN, N. WANG: "Diffraction by a Perfectly Conducting Rectangular Cylinder which is Illuminated by an Array of Line Sources", Report 3001-7, Electro-Science Laboratory, The Ohio State Univ., Columbus, Ohio (1973).
6.48. L.L. TSAI, D.R. WILTON, M.G. HARRISON, E.H. WRIGHT: IEEE Trans. Antennas Propagation AP-**20**, 705 (1972).
6.49. J.B. KELLER, D.S. AHLUWALIA: SIAM J. Appl. Math. **20**, 390 (1971).
6.50. R.B. DYBDAL, R.C. RUDDUCK, L.L. TSAI: IEEE Trans. Antennas Propagation AP-**14**, 574 (1966).
6.51. R.C. RUDDUCK, L.L. TSAI: IEEE Trans Antennas Propagation AP-**16**, 84 (1968).
6.52. C.E. RYAN, JR., R.C. RUDDUCK: IEEE Trans. Antennas Propagation AP-**16**, 490 (1968).
6.53. R.C. RUDDUCK, D.C.F. WU: IEEE Trans. Antennas Propagation AP-**17**, 797 (1969).
6.54. H.Y. YEE, L.B. FELSEN, J.B. KELLER: SIAM J. Appl. Math. **16**, 268 (1968).
6.55. L.B. FELSEN, H.Y. YEE: IEEE Trans. Antennas Propagation AP-**16**, 268 and 360 (1968).
6.56. J.S. YU, R.C. RUDDUCK, L. PETERS, JR.: IEEE Trans. Antennas Propagation AP-**14**, 138 (1966).
6.57. J.S. YU, R.C. RUDDUCK: IEEE Trans. Antennas Propagation AP-**17**, 651 (1969).
6.58. C.A. MENTZER, L. PETERS, JR., R.C. RUDDUCK: Submitted for publication.

6.59. T. Hua: "The Reflection Coefficient of a Horn-Waveguide Junction", M.Sc. Thesis, The Ohio State Univ., Columbus, Ohio (1970).
6.60. M. A. K. Hamid: IEEE Trans. Antennas Propagation AP-**16**, 520 (1968).
6.61. D. L. Hutchins, R. G. Kouyoumjian: J. Acoust. Soc. Am. **45**, 485 (1969).
6.62. W. D. Burnside, R. J. Marhefka, C. L. Yu: "Roll Plane Analysis of On-Aircraft Antennas" Pre-print 139 (AGARD Conference on Antennas for Avionics, Munich, Germany, 1973).
6.63. F. Molinet, L. Saltiel: "High Frequency Radiation Pattern Prediction for Satellite Antennas", Final Report ESTEC Contract 1820/72 HP, Laboratoire Central de Telecommunications, Velizy-Villacoublay, France (1973).
6.64. J. Bach, K. Pontoppidan, L. Solymar: "High Frequency Radiation Pattern Prediction for Satellite Antennas", Final Report ESTEC Contract 1821/72 HP, The Technical University of Denmark, Lyngby (1973).
6.65. B. D. Seckler, J. B. Keller: J. Acoust. Soc. Am. **31**, 192 (1959).
6.66. R. M. Lewis, N. Bleistein, D. Ludwig: Comm. Pure Appl. Math. **20**, 295 (1967).
6.67. B. Rulf, L. B. Felsen: In *Quasi-Optics*, ed. by J. Fox (Polytechnic Press, Brooklyn, 1964), pp. 107–149.
6.68. H. M. Nussenzveig: In *Methods and Problems of Theoretical Physics*, ed. by J. E. Bowcock (North Holland Publishing Co., Amsterdam, 1970).
6.69. L. B. Felsen, N. Marcuvitz: *Radiation and Scattering of Waves* (Prentice-Hall, Inc., Englewood Cliffs, N.J., 1973).
6.70. P. Ya. Ufimtsev: *Metod Krayevykh Voiln v Fizicheskoy Teorii Difraktsii* (Sovetskoye Radio, 1962). For an English translation see Foreign Technology Division, Document ID No. FTD-HC-23-259-71 (1971).
6.71. V. I. Ivanov: USSR Comput. Math. and Math. Phys. **2**, 216 (1971).

7. Reflector Antennas

W. V. T. RUSCH

With 24 Figures

Digital computers, when used properly, have achieved outstanding success in the analysis of reflector-antenna systems. Confidence in the accuracy of problem formulation and evaluation has achieved the point that experimental test data, previously considered the sacred be-all and end-all in antenna analysis, can be verified or replaced by the results of computer analysis. Under certain circumstances the accuracies obtainable in computer analysis exceed those which can be achieved by careful experimentation, and instances are not infrequent where unsuspected sources of experimental error are exposed by concurrent analysis. Furthermore, the time and cost factors of experimental versus numerical analysis are generally commensurate and frequently weighted in favor of the computer approach. In practice, a combination of experimental and computational programs appropriate to the nature of the problem is most desirable, with the computer used heavily in the design stages and the experimental program primarily oriented toward the final test and system verification.

The material of this chapter will present both background material and recent developments in the application of computers to the analysis of reflector antennas. The material will be applicable to most reflector contours, although primary attention will be given to focusing reflectors and focusing systems. Emphasis will be placed on the numerical techniques and procedures, rather than mathematical proofs and rigorous derivations of the fundamental equations of scattering and diffraction. On the other hand, specific examples involving numbers and technological details will usually be presented only to illustrate the numerical procedures. Consequently, large amounts of practical design data will not be presented. A reasonably representative bibliography is appended to these notes for the reader who wishes to pursue either mathematical rigor or practical design data in greater detail.

7.1. Formulation of the Field Equations

7.1.1. Free-Space Dyadic Green's Function

The electromagnetic characteristics of perfectly conducting reflector antennas may be analyzed with a variety of diffraction-theoretic techniques. Most of these techniques involve integrations over the currents induced on the reflector by an assumed *known* impressed field. The scattered fields radiated by these currents are, in terms of the geometry of Fig. 7.1 [7.1]

$$\boldsymbol{E}(P) = -\frac{j\omega\mu_0}{4\pi}\int_S \left[\boldsymbol{J}_S\psi + \frac{1}{k^2}(\boldsymbol{J}_S\cdot\nabla)\nabla\psi\right]dS\,, \tag{7.1}$$

$$\boldsymbol{H}(P) = \frac{1}{4\pi}\int_S [\boldsymbol{J}_S \times \nabla\psi]\,dS \tag{7.2}$$

where $\psi = \exp(-jkr)/r$. The operator ∇ within the field integrals operates only on the coordinates of the source point $\boldsymbol{\varrho}$. Consequently

$$\nabla\psi = (jk + 1/r)(\exp(-jkr)/r)\hat{\boldsymbol{a}}_r\,, \tag{7.3}$$

$$\begin{aligned}(\boldsymbol{J}_S\cdot\nabla)\nabla\psi = \Bigg[&-k^2(\boldsymbol{J}_S\cdot\hat{\boldsymbol{a}}_r)\hat{\boldsymbol{a}}_r + \frac{3}{r}\left(jk + \frac{1}{r}\right)(\boldsymbol{J}_S\cdot\hat{\boldsymbol{a}}_r)\hat{\boldsymbol{a}}_r \\ &-\boldsymbol{J}_S\left(j\frac{k}{r} + \frac{1}{r^2}\right)\Bigg](\exp(-jkr)/r)\,.\end{aligned} \tag{7.4}$$

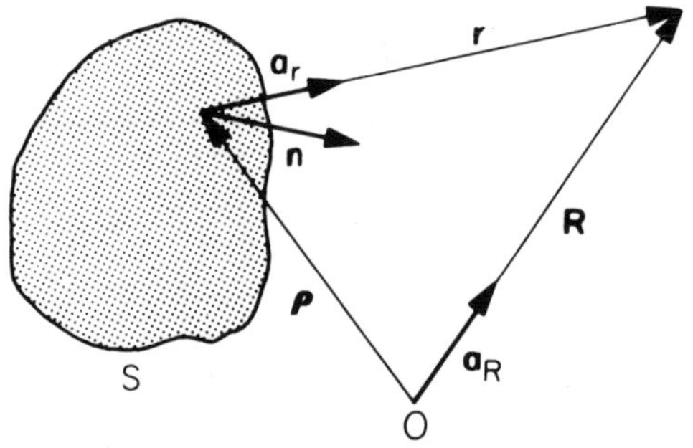

Fig. 7.1. Geometry of vectors for surface integrals

The integrand of (7.1) is frequently written in terms of a linear operator $\underset{\sim}{\boldsymbol{\Gamma}}$ known as the free-space dyadic Green's function [7.2]

$$\boldsymbol{E}(P) = -j\omega\mu_0 \int_S \underset{\sim}{\boldsymbol{\Gamma}} \cdot \boldsymbol{J}_S \, dS\,, \tag{7.5}$$

where the components of the dyadic $\underset{\sim}{\boldsymbol{\Gamma}}$ are found in (7.4). Near-field calculations, e.g., focal-region calculations, require that this most general form of $\underset{\sim}{\boldsymbol{\Gamma}}$ be used.

When $r \gg \lambda$, the leading term dominates (7.4) and the fields become

$$\boldsymbol{E}(P) = -\frac{j\omega\mu_0}{4\pi} \int_S [\boldsymbol{J}_S - (\boldsymbol{J}_S \cdot \hat{\boldsymbol{a}}_r)\hat{\boldsymbol{a}}_r]\, \psi \, dS\,, \tag{7.6}$$

$$\boldsymbol{H}(P) = \frac{1}{4\pi} \int_S [\boldsymbol{J}_S \times \nabla\psi] \, dS\,. \tag{7.7}$$

These somewhat simpler expressions are used in Fresnel-region calculations. When the field Point P is sufficiently far from the sources that the Fraunhofer approximations can be made, the fields at such distant points become

$$\boldsymbol{E}(P) = -\frac{j\omega\mu_0}{4\pi} \frac{\exp(-jkR)}{R} \int_S [\boldsymbol{J}_S - (\boldsymbol{J}_S \cdot \hat{\boldsymbol{a}}_R)\hat{\boldsymbol{a}}_R] \exp(jk\boldsymbol{\varrho} \cdot \hat{\boldsymbol{a}}_R) \, dS \tag{7.8}$$

$$\boldsymbol{H}(P) = \frac{jk}{4\pi} \frac{\exp(-jkR)}{R} \int_S [\boldsymbol{J}_S \times \hat{\boldsymbol{a}}_R] \exp(jk\boldsymbol{\varrho} \cdot \hat{\boldsymbol{a}}_R) \, dS\,. \tag{7.9}$$

Equations (7.8) and (7.9), considerably simplified by the fact that $\hat{\boldsymbol{a}}_R$ is constant with respect to the integration variables, are used for determining such far-field properties as gain and radiation pattern. Numerically, however, (7.1) and (7.6) are not a great deal more difficult to evaluate than (7.8).

7.1.2. Physical Optics

If the induced surface-current distributions are known, evaluation of the fields becomes straightforward, although the computations may be lengthy and laborious. However, only in a few special cases, e.g., sphere, ellipsoid, etc., can the currents be determined rigorously. The method-of-moments can be used to determine induced currents on small or moderately sized scatterers; however, the applicability of this technique to large focusing reflectors may be limited by computation cost and accuracy. One special two-dimensional example of the solution of an integral equation for a large reflector is treated later in this chapter.

In the event that the integrals have stationary points, the fields can be evaluated in terms of one or more simple, closed-form expressions which frequently lend themselves to greatly simplified geometrical ray-tracing interpretations. Examples of these geometrical techniques are presented in a subsequent section.

In the event that focused or nearly focused conditions obtain, however, the entire reflector is part of a single Fresnel zone. An obvious example is the main beam of the radiation pattern of a paraboloid reflector. Simple, localized stationary points do not exist in the field integrals. It is then necessary to evaluate contributions from all parts of the reflector. A technique that has found wide acceptance under those conditions is known in scattering theory as "Physical Optics" [7.3–5]. Physical Optics (PO) simply approximates the currents on the reflector by the currents calculated from the theory of Geometrical Optics (GO). These approximate current distributions are then used in (7.1)–(7.9) to determine the scattered field.

No rigorous justification for the PO approximations has been established. On the contrary, it can be shown [7.5] that PO in general fails to satisfy the reciprocity theorem everywhere except in the direction of a specular return [1]. In spite of this and other shortcomings, PO is an approximation technique that has proven very successful in the analysis of large reflector antennas, particular under focused or nearly focused conditions.

7.1.3. Aperture Formulation

In the event that the antenna configuration has a well-defined aperture, e.g. focusing reflectors, horns, and arrays, an equivalent aperture formulation may be used to compute the scattered field [7.6]

$$\begin{aligned} \boldsymbol{E}(P) = -\frac{j\omega\mu_0}{4\pi}\left(\frac{\exp(-jkR)}{R}\right) \int_{\text{aperture}} \Big\{ &-\frac{1}{\eta}(\hat{\boldsymbol{n}}\times\boldsymbol{E})\times\hat{\boldsymbol{a}}_{\mathrm{R}} \\ &+[\hat{\boldsymbol{n}}\times\boldsymbol{H}-((\hat{\boldsymbol{n}}\times\boldsymbol{H})\cdot\hat{\boldsymbol{a}}_{\mathrm{R}})\hat{\boldsymbol{a}}_{\mathrm{R}}]\Big\} \exp(jk\boldsymbol{\varrho}\cdot\hat{\boldsymbol{a}}_{\mathrm{R}})\,dS\,. \end{aligned} \tag{7.10}$$

The aperture fields must be approximated using, for example, geometrical optics, known unperturbed waveguide fields, etc. In many circumstances (7.10) is more appropriate for solution than the surface integrals. In the case of the paraboloid, at boresight the PO surface integral of (7.8) and

[1] It may also be shown that PO satisfies reciprocity for a focused paraboloid and a distant axial field point.

the aperture integral of (7.10) are equivalent. In directions far from the reflector axis, however, the aperture formulation introduces a significant path-length discrepancy [7.7].

7.2. Numerical Integration Procedures

The field integrals of Section 7.1 can be carried out analytically only in rare cases. Generally, because of the complexity of the integrand or because the integrand can be expressed only in terms of tabulated empirical data, such integrals must be numerically evaluated by machine computation. In some cases, which will be discussed below, the field expressions can be reduced to a one-dimensional integral; however, the remaining one-dimensional expressions must usually be evaluated numerically.

In general, the field integrals are of the form

$$\begin{aligned} \boldsymbol{E}(R,\theta,\phi) = \iint & \boldsymbol{F}[R,\theta,\phi,\varrho(\theta',\phi'),\theta',\phi'] \\ & \times \exp\{jk\gamma[R,\theta,\phi,\varrho(\theta',\phi'),\theta',\phi']\}\, d\theta'\, d\phi' \end{aligned} \tag{7.11}$$

or alternatively

$$\begin{aligned} \boldsymbol{E}(x,y,z) = \iint & \boldsymbol{F}[x,y,z,x',y',z'(x',y')] \\ & \times \exp\{jk\gamma[x,y,z,x',y',z'(x',y')]\}\, dx'\, dy', \end{aligned} \tag{7.12}$$

where (R, θ, ϕ) or (x, y, z) are the coordinates of the field point; $(\varrho, \theta', \phi')$ or (x', y', z') are the coordinates of the integration point on the surface or in the aperture; $\boldsymbol{E}$ and $\boldsymbol{F}$ are complex vector functions. A variety of integration schemes is available: trapezoidal integration, Simpson's rule, Gaussian quadratures, etc., each of which has advantages and limitations related ultimately to the nature of the integrand, the desired accuracy, and the available computer time. Fast-Fourier-transform techniques [7.8] and related algorithms may be applied to a limited class of scalar aperture integrals. However, such procedures have unfortunately not found wide applicability to the more complex vector diffraction integrals.

In the event that no characteristics of the integrand can be found that will permit simplification of the integral, it is often necessary to employ a "brute force" procedure to integrate over the entire surface or aperture. Examples of such situations are:

a) Focal-region and beam-scanning studies for a paraboloid reflector [7.9].

b) Radiation from and focal-region studies for a horn-reflector antenna [7.10].

c) Blocking of an aperture antenna by an offset circular obstacle.

d) Coupling between two Fresnel-region coaxial paraboloidal reflectors.

7.2.1. Physical-Optics Analysis of Radiation from a General Surface of Revolution

Restrictions of the reflector surface to a truncated surface of revolution, i.e., a reflector with an axis of symmetry, limits the generality of geometries that can be analyzed. However, it enables the azimuthal part of the two-dimensional physical-optics surface integral to be carried out analytically. These analytical manipulations introduce more complicated functions into the integrand of the remaining integral, but the net computer time is reduced typically by a factor of 50 from the time required for an equivalent two-dimensional integration for the same reflector. Furthermore, axisymmetric surfaces encompass a wide class of reflector profiles used in current antenna practice: the paraboloid, the paraboloid with axisymmetric distortions, the disc, the hyperboloid, the ellipsoid, the sphere, etc. In terms of the coordinates of Fig 7.2 the axially symmetric reflector can be described by the general polar equation

$$k\varrho(\theta') = -1/g(\theta') \quad \text{for} \quad \theta_0 \leqq \theta' \leqq \pi\,, \tag{7.13}$$

where, for example, for a paraboloid of focal length f

$$g(\theta') = \frac{1-\cos\theta'}{4\pi(f/\lambda)} \tag{7.14}$$

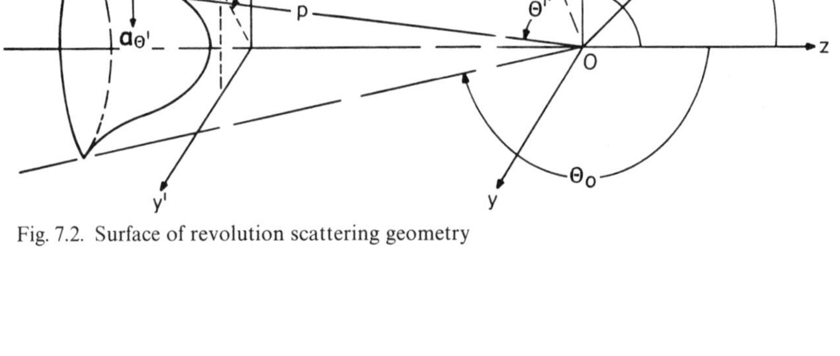

Fig. 7.2. Surface of revolution scattering geometry

and for a hyperboloid of eccentricity e

$$g(\theta') = \frac{1 + e\cos\theta'}{kep}. \tag{7.15}$$

Axisymmetric surfaces which cannot be described analytically can be determined experimentally and treated as a tabular function.

The incident electric field $\boldsymbol{E}_{\text{inc}}$ due to localized sources in the vicinity of O may be described by

$$\boldsymbol{E}_{\text{inc}} = \frac{\exp(-jk\varrho)}{\varrho} \left\{ \sum_{m=1}^{\infty} [a_m(\theta')\sin m\phi' + b_m(\theta')\cos m\phi']\,\hat{\boldsymbol{a}}_{\theta'} + \sum_{m=1}^{\infty} [c_m(\theta')\sin m\phi' + d_m(\theta')\cos m\phi']\,\hat{\boldsymbol{a}}_{\phi'} \right\}, \tag{7.16}$$

$$\boldsymbol{H}_{\text{inc}} = \frac{1}{\eta}\,\hat{\boldsymbol{a}}_{\varrho} \times \boldsymbol{E}_{\text{inc}}. \tag{7.17}$$

These Fourier series represent an asymptotic solution of the electromagnetic field equations, and, consequently, are capable of describing most far-zone fields. The Fourier coefficients are, in general, complex, thus encompassing linearly, circularly, or elliptically polarized feeds, feed-system phase errors, etc. These equations are particularly convenient for conical feed-horns with circular apertures. In the special case that $m = 1$ and only a_1 and d_1 are non-zero, for example, the feed is linearly polarized with the YZ-plane as the E plane

$$\boldsymbol{E}_{\text{inc}} = \frac{\exp(-jk\varrho)}{\varrho} \{a_1(\theta')\cos\phi'\,\hat{\boldsymbol{a}}_{\theta'} + d_1(\theta')\sin\phi'\,\hat{\boldsymbol{a}}_{\phi'}\}, \tag{7.18}$$

$$a_1(\pi) = -d_1(\pi). \tag{7.19}$$

The far-field scattering integrals in conjunction with the GO approximation for the induced surface currents then yield the scattered field. When the integration is carried out over an axially symmetric surface, the azimuthal integral can be carried out in closed form with varying degrees of approximation, depending upon the distance from the scatterer and the angle of observation. In the far field, at distances much greater than the maximum transverse dimension of the reflecting surface, the scattered field is given by

$$\boldsymbol{E}_\mathrm{S} = \frac{\exp(-jkR)}{R}\left\{\sum_m \left[f_m(\theta)\sin m\phi + g_m(\theta)\cos m\phi\right]\hat{\boldsymbol{a}}_\theta + \sum_m \left[h_m(\theta)\sin m\phi + k_m(\theta)\cos m\phi\right]\hat{\boldsymbol{a}}_\phi\right\}, \tag{7.20}$$

where

$$\begin{aligned} f_m(\theta) = \left(-\frac{1}{2}\right)(j)^m \int_{\theta_0}^{\pi} \frac{d\theta'\,\mathrm{e}^{-j\alpha}\sin\theta'}{[g(\theta')]^2} &\{a_m(\theta')\left[\cos\theta[g'(\theta')\sin\theta'\right.\\ &- g(\theta')\cos\theta']\,[J_{m-1}(\beta) - J_{m+1}(\beta)] - 2j[g'(\theta')\cos\theta'\\ &+ g(\theta')\sin\theta']\sin\theta\, J_m(\beta)]\\ &+ d_m(\theta')\,[-g(\theta')]\cos\theta[J_{m-1}(\beta) - J_{m+1}(\beta)]\}, \end{aligned} \tag{7.21}$$

$$\begin{aligned} g_m(\theta) = \left(-\frac{1}{2}\right)(j)^m \int_{\theta_0}^{\pi} \frac{d\theta'\,\mathrm{e}^{-j\alpha}\sin\theta'}{[g(\theta')]^2} &\{b_m(\theta')(\cos\theta[g'(\theta')\sin\theta'\\ &- g(\theta')\cos\theta']\,[J_{m-1}(\beta) - J_{m+1}(\beta)] - 2j[g'(\theta')\cos\theta'\\ &+ g(\theta')\sin\theta']\sin\theta\, J_m(\beta))\\ &+ c_m(\theta')\,g(\theta')\cos\theta[J_{m-1}(\beta) + J_{m+1}(\beta)]\}, \end{aligned} \tag{7.22}$$

$$\begin{aligned} h_m(\theta) = \left(\frac{1}{2}\right)(j)^m \int_{\theta_0}^{\pi} \frac{d\theta'\,\mathrm{e}^{-j\alpha}\sin\theta'}{[g(\theta')]^2} &\{b_m(\theta')\,[g'(\theta')\sin\theta'\\ &- g(\theta')\cos\theta']\,[J_{m-1}(\beta) + J_{m+1}(\beta)]\\ &+ c_m(\theta')\,g(\theta')\,[J_{m-1}(\beta) - J_{m+1}(\beta)]\}, \end{aligned} \tag{7.23}$$

$$\begin{aligned} k_m(\theta) = \left(-\frac{1}{2}\right)(j)^m \int_{\theta_0}^{\pi} \frac{d\theta'\,e^{-j\alpha}\sin\theta'}{[g(\theta')]^2} &\{a_m(\theta')\,[g'(\theta')\sin\theta'\\ &- g(\theta')\cos\theta']\,[J_{m-1}(\beta) + J_{m+1}(\beta)]\\ &- d_m(\theta')\,g(\theta')\,[J_{m-1}(\beta) - J_{m+1}(\beta)]\}, \end{aligned} \tag{7.24}$$

$$\alpha = \frac{\cos\theta'\cos\theta - 1}{g(\theta')}, \tag{7.25}$$

$$\beta = \frac{-\sin\theta\sin\theta'}{g(\theta')}. \tag{7.26}$$

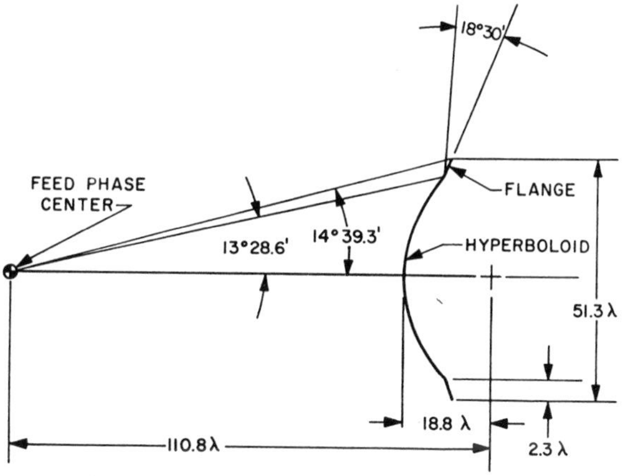

Fig. 7.3. Subreflector configuration for experimental scattered patterns

In particular, the transverse components of the far-zone field scattered from a paraboloid immersed in a linearly polarized field described by (7.18) and (7.19) are

$$E_\theta(P) = jkF \sin\phi \frac{\exp(-jkR)}{R} \int_{\theta_0}^{\pi} \frac{\exp[-jk\varrho(1-\cos\theta\cos\theta')]}{(1-\cos\theta')} \cdot \left\{a_1 \cos\theta[J_0(\beta) - J_2(\beta)] - d_1 \cos\theta[J_0(\beta) + J_2(\beta)] - 2j \sin\theta \operatorname{ctn}\frac{\theta'}{2} J_1(\beta) a_1\right\} \sin\theta' \, d\theta' , \quad (7.27)$$

$$E_\phi(P) = jkF \cos\phi \frac{\exp(-jkR)}{R} \int_{\theta_0}^{\pi} \frac{\exp[-jk\varrho(1-\cos\theta\cos\theta')]}{(1-\cos\theta')} \cdot \{a_1[J_0(\beta) + J_2(\beta)] - d_1[J_0(\beta) - J_2(\beta)]\} \sin\theta' \, d\theta' , \quad (7.28)$$

where $\beta = k\varrho \sin\theta \sin\theta'$.

To illustrate the generally excellent agreement between computed and measured patterns, the scattered field was determined for the reflector configuration shown in Fig. 7.3. This configuration is a 51.3-wavelength-diameter Cassegrainian subreflector consisting of a hyperboloid with a conical flange extension. The amplitude and phase patterns of the

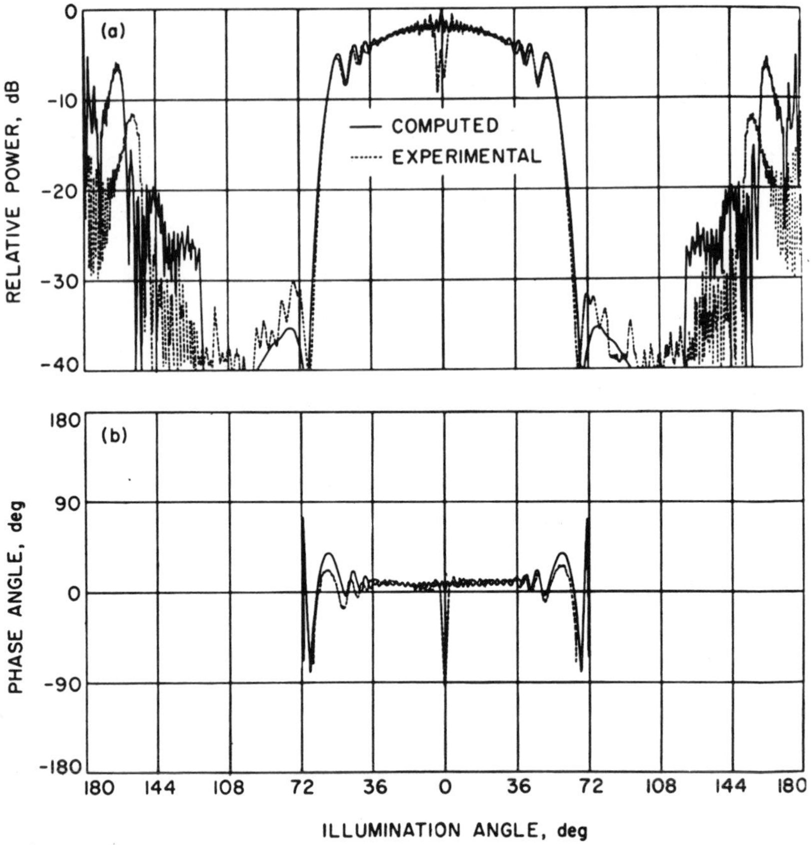

Fig. 7.4a and b. Comparison of computed and experimental amplitude and phase patterns

feedhorn used to illuminate the subreflector were measured on a 16-GHz antenna range, and these patterns were used as tabular input for the field calculations. The computed and experimental scattered-field amplitude and phase patterns (E-plane) are plotted in Fig. 7.4. Close agreement between the calculated and experimental pattern is evident, except for discrepancies near zero degrees, due to unavoidable feedhorn blockage of the experimental pattern, and at wide angles, were the reduced amplitude is more significantly affected by "parasitic" errors caused by ground reflections, inexact feedhorn pattern determination, etc. In the angular range from 140 to 180 degrees parallax in the relatively short antenna range also introduced errors. However, the field integrals based on the GO approximation for the current distribution have questionable validity when extrapolated into the rear hemisphere.

Convergence — An Example Using Dipole Illumination

The major restriction imposed by computer evaluation of the integrals is the machine time, which is a function of the required accuracy, the reflector size, and the number of observation points. Under many circumstances, for normal illumination tapers 0.1 dB accuracy requires a minimum of several integration steps per linear wavelength of the surface of integration. A more precise criterion for convergence, however, is the number of integration steps *per Fresnel zone*.

A numerical example has been worked out below for an illumination function that can also be integrated in closed form. The fields of a y-oriented, infinitesimal electric dipole can be represented by (7.18) with $a_1(\theta') = -\cos(\theta')$ and $d_1(\theta') = -1$. Using (7.27) or (7.28) the boresight field radiated from a paraboloid with this dipole illumination function is

$$E(R, \theta = 0) = -jkf \exp(-j2kf)(1 + \cos\theta_0)\frac{\exp(-jkR)}{R}. \qquad (7.29)$$

This field has been evaluated numerically from (7.28) using a one-dimensional Simpson's algorithm for the θ'-integration. The parameters used were $D/\lambda = 25$, $f/D = 0.4$, for which the closed-form result is

$$E(R, \theta = 0) = -j\,35.2988\frac{\exp(-jkR)}{R}. \qquad (7.30)$$

The numerical-integration results are tabulated in Table 7.1 as a function of the number of integration steps. Six-figure accuracy is achieved with

Table 7.1. Calculated field magnitude as a function of number of integration steps

Number of polar integration steps	Field magnitude		
	$\theta = 0°$	$\theta = 3.75°$	$\theta = 6.15°$
2	35.3186	1.6299	10.1080
4	35.3000	4.7049	2.8796
8	35.2989	4.6764	2.3264
16	35.2988	4.6722	2.2817
32	35.2988	4.6719	2.2791
64	35.2988	4.6719	2.2789
128	35.2988	4.6719	2.2789

16 steps, although an accuracy of about 2 parts in 3500 is achieved with only 2 integration steps. In the case of a distant point on axis, the entire aperture constitutes a single Fresnel zone, and consequently only a few integration steps are needed for the 25-wavelength-diameter aperture. Results are also tabulated at $\theta = 3.75$ degree, the first H-plane sidelobe peak, for which comparable accuracy is achieved with 32 steps, and at $\theta = 6.15$ degree, the second H-plane sidelobe peak, for which comparable accuracy is achieved with 64 steps. Two general comments can be made:

a) The proper number of integration steps to achieve a desired accuracy is primarily determined by the rate at which the phase of the integrand changes over the integration interval and consequently is a function of (among other factors) the angle of observation.

b) One can frequently achieve a reasonable accuracy with a surprisingly small number of integration steps.

Interval-Halving with Automatic Testing

Classical quadrature algorithms generally require that the number of integration steps be preprogrammed. However, if accuracy is desired over a wide angular range, one must generally be too conservative in specifying the number of integration steps beforehand. Usually 5–10 steps per linear wavelength are chosen, independent of the observation angle, resulting in "overkill" when the aperture is near to focused conditions.

On the other hand, using a standard trapezoid or Simpson's algorithm with successive interval halving until a preprogrammed *accuracy* has been achieved permits the computer to "select" the appropriate number of integration steps. Such automated algorithms also minimize the number of integrand evaluations by making use of all previous evaluations at every stage without repeating them. Further accuracy for a given number of integrand evaluations can frequently be achieved by extrapolating to zero-grid-size (Romberg's integration) using a sequence of successively halved trapezoid results [7.11].

The advantages of Romberg's quadrature are illustrated by the following example: a prime-focus paraboloid reflector is 73.2 wavelengths in diameter; the angular distribution of the field from the feed is $\cos^2(\pi - \theta')$; and the desired accuracy is 1% of the field in a particular direction or 0.01% of the field in the direction of the main beam, whichever is achieved first. For these parameters, only 8 integration steps in the θ'-integration were required to compute the boresight field. However, 512 integration steps were required to compute the field in a direction 40 beamwidths off-axis, or about 14 steps per linear wavelength. The machine itself determines the proper number of integration steps for each angle of observation.

An extensive literature exists for one- and two-dimensional Romberg-type numerical integration algorithms, most of which are applicable to the problem of numerical evaluation of the field integrals. Limitations of space prevent a more extensive discussion here.

7.2.2. Ludwig Algorithm

In the event that the field integral cannot be reduced analytically, LUDWIG [7.12] has developed a two-dimensional method conceptually similar to Filon's method that achieves a significant reduction in computer time. The amplitude and phase of the integrand are expanded in a linearized Taylor series for each differential surface element at (θ'_m, ϕ'_n) of the order of a square wavelength in size. Thus, with reference to (7.11) and (7.12)

$$F \cong a_{mn} + b_{mn}(\theta' - \theta'_m) + c_{mn}(\phi' - \phi'_n), \tag{7.31}$$

$$\gamma \cong \alpha_{mn} + \beta_{mn}(\theta' - \theta'_m) + \zeta_{mn}(\phi' - \phi'_n). \tag{7.32}$$

The resulting expression can be integrated in closed form for each differential surface element. With incremental areas two-thirds of a square wavelength, for example, absolute errors are more than 40 dB below the main field maximum. Such accuracy can be achieved on a straightforward two-dimensional integration using Simpson's rule only if the differential areas are at least 16 times smaller. The total reduction in computer time using Ludwig's technique is a factor of 4–8.

7.3. High-Frequency Reflectors with Stationary Points

In the evaluation of the field integrals particular attention should be given to regions where the overall phase function of the integrand is stationary. Should these stationary points occur, the fields can be evaluated asymptotically in terms of simple closed-form expressions. Stationary points of the first kind satisfy the condition

$$\frac{\partial \gamma}{\partial x'} = \frac{\partial \gamma}{\partial y'} = 0 \tag{7.33}$$

or, if the boundary of the aperture or open surface is specified by $G(x', y') = 0$, stationary points of the second kind are determined by the condition [7.13, 14]

$$\frac{\partial G}{\partial x'} \frac{\partial \gamma}{\partial y'} - \frac{\partial G}{\partial y'} \frac{\partial \gamma}{\partial x'} = 0. \tag{7.34}$$

In the limit of large k (short wavelength) major contributions to the field come from the immediate vicinity of these stationary points, provided that the magnitude of the integrand does not vanish at such points. Standard saddle-point techniques can be used to evaluate these asymptotic contributions.

7.3.1. Shadow-Region Cancellation

Stationary points of the first kind lie at points on the reflector intersected by the straight line from the feed to the field point. Thus these points exist only when P lies within the shadow "cast" by the reflector when it is illuminated by the incident field. In the limit $k \to \infty$, the scattered field exactly cancels the incident field behind the reflector, producing, as expected, a perfect geometrical shadow in that region. When P lies *along* the shadow boundary, the stationary point lies on the edge of the reflector, thus creating only "half" a saddle point. At such points only half of the incident field is cancelled, thus producing the classical result of a -6 dB field at the edge of the shadow.

LUDWIG [7.15] has suggested an ingenious feed synthesis procedure which he explains in terms of the physical-optics scattering integrals. However, for large reflectors the technique is readily interpretable in terms of shadow-region cancellation.

a) Assume that it is necessary to synthesize a given vector field $\boldsymbol{E}_S$ with a given reflector. The problem of determining an appropriate feed pattern may be solved by permitting $\boldsymbol{E}_S$ to illuminate the back (wrong) side of the reflector, inducing an appropriate current distribution $\boldsymbol{J}_1$ shown in Fig. 7.5a. This current density radiates directively, producing an easily calculable field $\boldsymbol{E}_F$ behind the reflector but, in the short wavelength limit, producing $-\boldsymbol{E}_S$ in front of (to the left of) the reflector in order to produce zero total field there.

b) The known field $\boldsymbol{E}_F$ is then used to illuminate the front (left) of the reflector, inducing current distribution $\boldsymbol{J}_2$ shown in Fig. 7.5b. This current density produces $-\boldsymbol{E}_F$ behind the reflector, since the total field in this shadow region must be zero. Since J_1 produces $\boldsymbol{E}_F$ to the right and $\boldsymbol{J}_2$ produces $-\boldsymbol{E}_F$ to the right, we conclude that

$$\boldsymbol{J}_2 = -\boldsymbol{J}_1 \tag{7.35}$$

and, consequently, $\boldsymbol{J}_2$ produces $+\boldsymbol{E}_S$ to the left. Thus the desired field $\boldsymbol{E}_S$ has been achieved by using the feed function $\boldsymbol{E}_F$ to illuminate the front of the reflector. $\boldsymbol{E}_F$, in turn, has been determined by allowing $\boldsymbol{E}_S$ to illuminate the wrong side of the reflector!

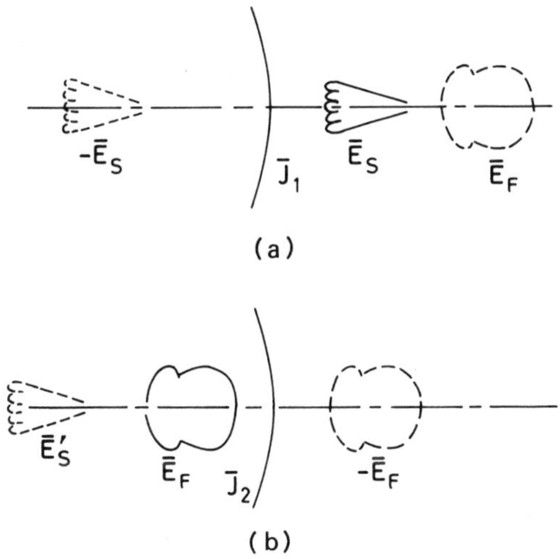

Fig. 7.5a and b. Feed synthesis procedure based on shadow-region cancellation

7.3.2. Geometrical Optics

Additional stationary points of the first kind for non-focused reflectors correspond to the classical geometrical-optics ray. The geometrical optics of reflectors in free space is treated elsewhere in the literature [7.18]. The reflected field $\boldsymbol{E}_r(\mathrm{P})$ at a Point P along a reflected ray is related to the reflected field $\boldsymbol{E}_r(\mathrm{S})$ at the point of reflection by

$$\boldsymbol{E}_r(\mathrm{P}) = \boldsymbol{E}_r(\mathrm{S}) \sqrt{\frac{|\beta_1|\,|\beta_2|}{|R_{\mathrm{SP}} - \beta_1|\,|R_{\mathrm{SP}} - \beta_2|}} \exp(j\delta_c) \exp(-jkR_{\mathrm{SP}}), \quad (7.36)$$

where

R_{SP} = the distance from S to P.

β_1, β_2 = principal radii of curvature (caustic distances) of the tube of rays reflected from S, positive if the center of curvature lies on the same side of the reflector as P, negative if the center of curvature lies on the opposite side of the reflector.

$$\delta_c = \begin{cases} 0, & \text{if neither center of curvature lies between S and P.} \\ \dfrac{\pi}{2}, & \text{if one center of curvature lies between S and P.} \\ \pi, & \text{if both centers of curvature lie between S and P.} \end{cases}$$

For a reflecting surface of revolution r', the distance from S, a point on the surface, to the z-axis is a function of z', the axial coordinate, but not of ϕ', the azimuthal coordinate. For such a surface the caustic distances, β_1 and β_2, are roots of the quadratic equation [7.17]

$$(EG-F^2)\beta^2+(Eg-2Ff+Ge)\beta+(eg-f^2)=0\,, \tag{7.37}$$

where

$$E=\frac{\partial\hat{s}_r}{\partial r'}\cdot\frac{\partial\hat{s}_r}{\partial r'}\,,\qquad e=\frac{\partial \boldsymbol{R}}{\partial r'}\cdot\frac{\partial\hat{s}_r}{\partial r'}\,,$$

$$F=\frac{\partial\hat{s}_r}{\partial r'}\cdot\frac{\partial\hat{s}_r}{\partial \phi'}\,,\qquad f=\frac{\partial \boldsymbol{R}}{\partial \phi'}\cdot\frac{\partial\hat{s}_r}{\partial r'}\,,$$

$$G=\frac{\partial\hat{s}_r}{\partial \phi'}\cdot\frac{\partial\hat{s}_r}{\partial \phi'}\,,\qquad g=\frac{\partial \boldsymbol{R}}{\partial \phi'}\cdot\frac{\partial\hat{s}_r}{\partial \phi'}\,,$$

and $\hat{s}_r$ is the reflected wave-normal, and $\boldsymbol{R}$ is the position vector from the origin to the reflection Point S. For a point-source feed located on the z-axis, both F and f reduce to zero, yielding

$$\beta_1=-e/E\,, \tag{7.38a}$$

$$\beta_2=-g/G\,. \tag{7.38b}$$

Geometrical optics may be used for focal-region studies of a receive-mode paraboloid [7.18–20] in order to locate the general position of the

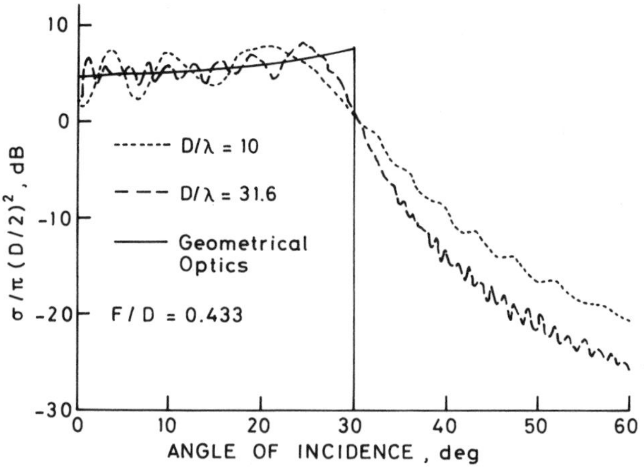

Fig. 7.6. Normalized radar cross-section of a paraboloid

most intense focal-region fields and to grossly characterize these fields. However, the classical geometrical-optics analysis is incapable of yielding useful intensity information on the focal-region fields.

The high-frequency radar cross-section of a reflector is

$$\sigma = 4\pi |\beta_1| \, |\beta_2| \, , \tag{7.39}$$

where β_1 and β_2 are functions of the angle of incidence and reflector geometry. For a paraboloid with an axially incident plane wave, $|\beta_1| = |\beta_2| = f$. The normalized radar cross-section of a paraboloid is plotted in Fig. 7.6 as a function of the angle of incidence, together with PO results for $D/\lambda = 10$ and $D/\lambda = 31.6$.

The basic properties of the Cassegrain dual-reflector antenna are based on principles of ray optics: the subreflector is placed so that its foci coincide with the focus of the paraboloid reflector and the phase center of the primary feed. A spherical wave from the primary feed will be transformed by the subreflector into a spherical wave emerging from the focus of the paraboloid; this wave will then be reflected by the paraboloid and radiated into space. Ray tracing indicates no spillover beyond the edge of the paraboloid and, consequently a low effective noise temperature since radiations will not be received from the ground.

Some recent large dual reflector systems [7.21] have been built after the principles of the Gregorian telescope using a concave ellipsoidal subreflector placed beyond the focus of the paraboloidal so that its two foci coincide with the paraboloid focus and the primary feed phase center. Every ray emerging from the primary feed and reflected by the ellipsoid will converge toward the paraboloid focus and then diverge toward the paraboloid. The ray behavior is similar to the Cassegrain system except for "inversion" and longer ray paths.

Geometrical ray tracing reveals that any combination of reflectors having confocal conic sections will produce the same aperture field distribution as that of a single prime-focus paraboloidal reflector having the equivalent focal length of the compound system [7.22, 23]. Thus, as shown in Fig. 7.7, the combination of main dish and subdish may be considered as being replaced by an equivalent focusing surface at a certain distance from the real primary focus of a Cassegrain or Gregorian system. This surface is defined as the locus of intersection of incoming rays parallel to the antenna axis with the extension of the corresponding rays converging toward the real focal point. This equivalent relecting surface has a paraboloidal contour with a focal length equal to the distance from its vertex to the real focal point [7.24]. As a result, this surface could be employed as a reflecting dish which would focus an incoming plane wave toward the real focal point in exactly the same

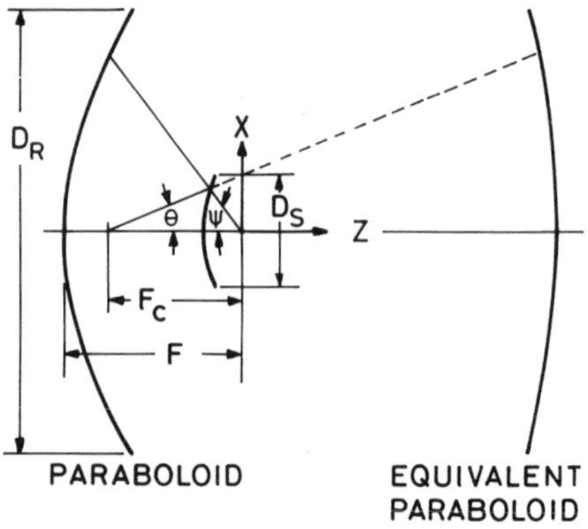

Fig. 7.7. Geometry of equivalent paraboloid

manner as does the combination of main dish and subdish. (Actually, the plane wave would have to be incident from the opposite direction.) Thus the analysis problem reduces to the considerably simpler analysis of a prime-focus system with the same feed and a paraboloidal reflector with the same diameter but a much larger focal length given by

$$F_{\text{equiv}} = \left(\frac{e+1}{e-1}\right) F\,, \tag{7.40}$$

where e is the subreflector ellipticity.

Geometrical-optics techniques have been developed [7.25–26] to determine the necessary dual-reflector combination which satisfies the boundary conditions imposed by an assumed primary-feed field distribution, and a desired field distribution $I(x)$ in the aperture of the modified paraboloid. These equations are based on the principles of geometrical optics:

a) At each reflector the incident and reflected rays and the surface-normal are coplanar, and the angle of incidence equals the angle of reflection (law of reflection).

b) Energy flow along each differential tube of rays remains constant, even when the tube undergoes reflection (conversation of energy).

c) Ray directions are normal to the constant-phase surfaces, and this condition is maintained after reflections (theorem of Malus).

Rearranging these three simultaneous differential equations yields a form suitable for numerical integration. Because the procedure is based upon geometrical optics, valid solutions are limited to surfaces with large radii of curvature. Furthermore, diffraction effects such as spillover are not included.

Maximum aperture efficiency is generally achieved with uniform illumination, i.e. setting $I(x)$ equal to a constant. The resulting significant increase in efficiency is evidenced by the performance of large operational systems. For example, a 90-foot Comsat dual-shaped-reflector system in Fucino, Italy, operates from 3.7 to 6.4 GHz with a reported overall efficiency of 74% measured at the input of the receiver [7.27]. Other distribution functions $I(x)$ may be used if different radiation characteristics such as ultra-low sidelobes are desired [7.28].

7.3.3. Stationary Points of the Second Kind (Edge Rays)

Stationary points of the second kind defined by (7.34) correspond to points on the reflector rim. For example, for any axially symmetric reflector with an on-axis feed, two stationary points of the second kind lie where the plane containing the reflector axis and the field point intersects the rim of the reflector. These ray-like contributors which appear to emerge from each stationary point may be expressed as the product of

a) the tangential incident-field component;

b) a phase factor corresponding to the path length from the stationary point to the field point;

c) a caustic divergence factor;

d) a diffraction coefficient.

The diffraction coefficient results from saddle-point evaluation of the physical-optics field integrals. However, the geometrical theory of diffraction, described in the following subsection, can be expected to provide improved diffraction coefficients because it more accurately models the reflector surface-current density in the vicinity of the edge.

7.3.4. Geometrical Theory of Diffraction for Reflectors in Free Space

The Geometrical Theory of Diffraction (GTD) developed in Chapter 6 [7.29] may be applied advantageously to many reflector antenna geometries. This subsection presents analytical, computational, and experimental results for commonly encountered reflector geometries, both to illustrate the general principles and to present a compact summary of generally applicable formulas.

Axially Symmetric Reflector for a Vector Spherical-Wave Point Source on Axis

If a point-source feed is located on the axis of an axially symmetric reflector, then only two singly edge-diffracted rays are possible for a

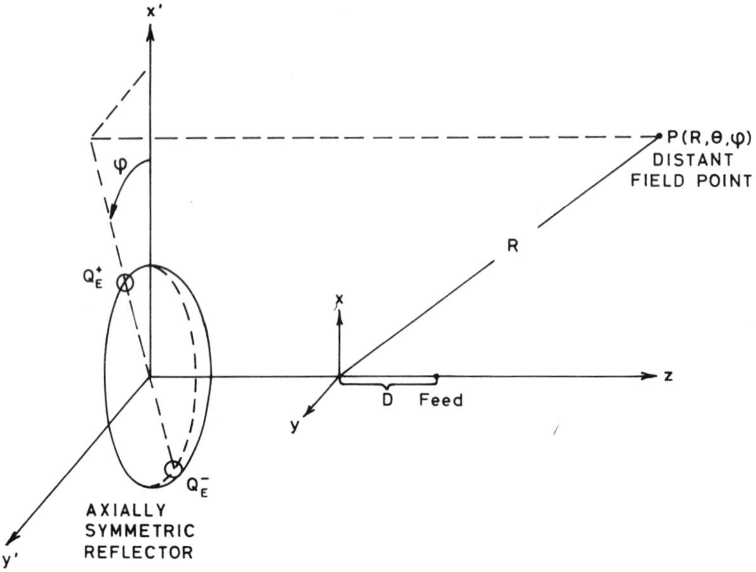

Fig. 7.8. Geometry of edge-diffraction points for an axially symmetric reflector with feed on axis

distant field point $P(R, \theta, \phi)$ if $\theta \neq 0, \pi$ (Fig. 7.8). These rays are "diffracted" from Q_E^+, where the plane $\phi = \phi$ intersects the edge on the *same* side of the z-axis as the field point, and from Q_E^-, where the plane $\phi = \phi + \pi$ intersects the edge diametrically opposite Q_E^+ on the opposite side of the z-axis from the field point. Q_E^+ and Q_E^- correspond exactly to the plus and minus stationary points from saddle-point theory (see above).

The geometry of the edge-diffracted ray from Q_E^- is shown in Fig. 7.9. (The surface is assumed to be convex.) The ray from the feed to Q_E^- defines θ'_{edge}, and the corresponding extreme geometrically reflected ray, when extended back to the z-axis, defines θ_{edge}. These two angles may then be used to define two intermediate angles used in the edge-diffraction notation of KELLER [7.30]:

$$\delta_t = \frac{\theta_{\text{edge}} - \theta'_{\text{edge}}}{2}, \tag{7.41a}$$

$$\alpha = \frac{\theta_{\text{edge}} + \theta'_{\text{edge}}}{2}. \tag{7.41b}$$

Let the feed radiation be defined by

$$\boldsymbol{E}_f = [E_{f\theta}(\theta, \phi)\hat{\boldsymbol{a}}_\theta + E_{f\phi}(\theta, \phi)\hat{\boldsymbol{a}}_\phi] \frac{\exp(-jkR)}{R} \tag{7.42}$$

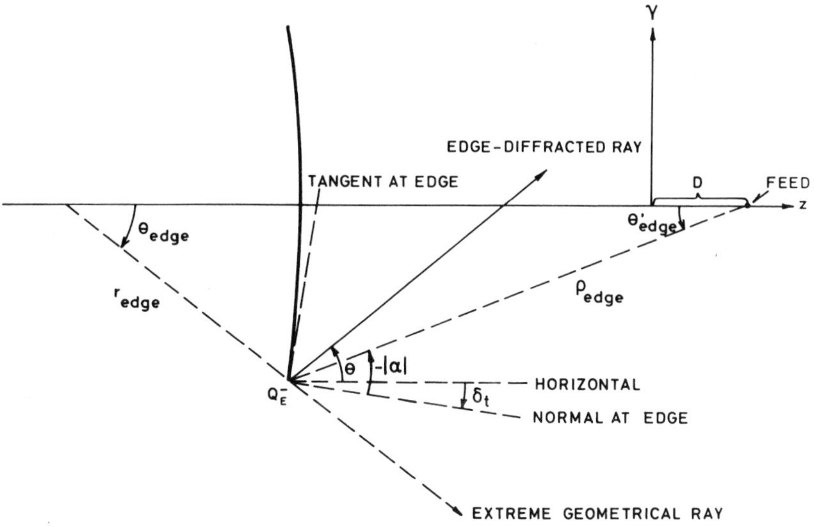

Fig. 7.9. Geometry of ray diffracted from Q_E^-

and the total singly edge-diffracted field be defined by

$$\boldsymbol{E}_\mathrm{d} = [E_{\mathrm{d}\theta}(\theta,\phi)\hat{\boldsymbol{a}}_\theta + E_{\mathrm{d}\phi}(\theta,\phi)\hat{\boldsymbol{a}}_\phi]\,\frac{\exp(-jkR)}{R}\,, \tag{7.43}$$

where the angles and unit vectors for both sets of fields are defined in the sense indicated in Fig. 7.8. Then, the components of $\boldsymbol{E}_\mathrm{d}$ can be decomposed into contributions from both Q_E^+ and Q_E^-

$$E_{\mathrm{d}\theta}(\theta,\phi) = E_{\mathrm{d}\theta}^+(\theta,\phi) + E_{\mathrm{d}\theta}^-(\theta,\phi)\,, \tag{7.44a}$$

$$E_{\mathrm{d}\phi}(\theta,\phi) = E_{\mathrm{d}\phi}^+(\theta,\phi) + E_{\mathrm{d}\phi}^-(\theta,\phi)\,. \tag{7.44b}$$

Considering first the diffracted field from Q_E^- (because no transition functions are required for these rays)

$$\begin{aligned} E_{\mathrm{d}\theta}^-(\theta,\phi) = &\left\{-\frac{E_{\mathrm{f}\theta}(\pi-\theta'_\mathrm{edge},\phi+\pi)\exp(-jk\varrho_\mathrm{edge})}{\varrho_\mathrm{edge}}\right\} \\ &\cdot\left\{\exp(j\pi/2)\sqrt{\frac{D_\mathrm{refl}/2}{\sin\theta}}\right\} \\ &\cdot\{\exp\{jk[(2c+Z_\mathrm{edge})\cos\theta-(D_\mathrm{refl}/2)\sin\theta]\}\}\,D_\mathrm{h}^-\,, \end{aligned} \tag{7.45}$$

$$E^{-}_{\mathrm{d}\phi}(\theta,\phi)=\left\{-\frac{E_{\mathrm{f}\phi}(\pi-\theta'_{\mathrm{edge}},\phi+\pi)\exp(-jk\varrho_{\mathrm{edge}})}{\varrho_{\mathrm{edge}}}\right\} \cdot\left\{\exp(j\pi/2)\sqrt{\frac{D_{\mathrm{refl}}/2}{\sin\theta}}\right\} \cdot\{\exp\{jk[(2c+Z_{\mathrm{edge}})\cos\theta-(D_{\mathrm{refl}}/2)\sin\theta]\}\}\,D^{-}_{\mathrm{s}}\,, \tag{7.46}$$

where ϱ_{edge} is the distance from the feed to Q^{-}_{E}, Z_{edge} is the Z-coordinate of Q^{-}_{E} ($Z_{\mathrm{edge}}<0$), D_{refl} is the reflector diameter, $Z=-2c$ is an arbitrary phase reference point on the negative Z-axis, and the diffraction coefficients are given by [cf. (6.32)]

$$D^{-}_{\mathrm{s;h}}=\begin{cases}\left\{-\dfrac{\exp(-j\pi/4)}{2\sqrt{2\pi k}}\left[\dfrac{1}{\cos\left(\dfrac{\theta+\delta_{\mathrm{t}}+\alpha}{2}\right)}\mp\dfrac{1}{\sin\left(\dfrac{\theta+\delta_{\mathrm{t}}-\alpha}{2}\right)}\right]\right\}, & 0<\theta\leqq\dfrac{\pi}{2}-\delta_{\mathrm{t}}\\ 0, \quad \dfrac{\pi}{2}-\delta_{\mathrm{t}}<\theta<\dfrac{\pi}{2} & \\ \left\{-\dfrac{\exp(-j\pi/4)}{2\sqrt{2\pi k}}\left[\dfrac{1}{\cos\left(\dfrac{\theta+\delta_{\mathrm{t}}+\alpha}{2}\right)}\pm\dfrac{1}{\sin\left(\dfrac{\theta+\delta_{\mathrm{t}}-\alpha}{2}\right)}\right]\right\}, & \dfrac{\pi}{2}\leqq\theta<\pi\,,\end{cases} \tag{7.47}$$

where the upper signs are taken for the soft (s) coefficient and the lower signs are taken for the hard (h) coefficient. Notice that D^{-}_{s} is zero in the direction of the surface tangent, $\theta=\pi/2-\delta$, but D^{-}_{h} is not. Slightly modified but principally the same diffraction coefficients for single edge-diffracted rays may be derived for concave surfaces.

The geometry of the ray diffracted from Q^{+}_{E} is illustrated in Fig. 7.10. The fields of this ray are given by

$$E^{+}_{\mathrm{d}\theta}(\theta,\phi)=\left[\frac{E_{\mathrm{f}\theta}(\pi-\theta'_{\mathrm{edge}},\phi)\exp(-jk\varrho_{\mathrm{edge}})}{\varrho_{\mathrm{edge}}}\right]\sqrt{\frac{D_{\mathrm{refl}}/2}{\sin\theta}} \cdot\{\exp\{jk[(2c+Z_{\mathrm{edge}})\cos\theta+(D_{\mathrm{refl}}/2)\sin\theta]\}\}\,D^{+}_{\mathrm{h}}\,, \tag{7.48}$$

$$E^{+}_{\mathrm{d}\phi}(\theta,\phi)=\left[\frac{E_{\mathrm{f}\phi}(\pi-\theta'_{\mathrm{edge}},\phi)\exp(-jk\varrho_{\mathrm{edge}})}{\varrho_{\mathrm{edge}}}\right]\sqrt{\frac{D_{\mathrm{refl}}/2}{\sin\theta}} \cdot\{\exp\{jk[(2c+Z_{\mathrm{edge}})\cos\theta+(D_{\mathrm{refl}}/2)\sin\theta]\}\}\,D^{+}_{\mathrm{s}}\,, \tag{7.49}$$

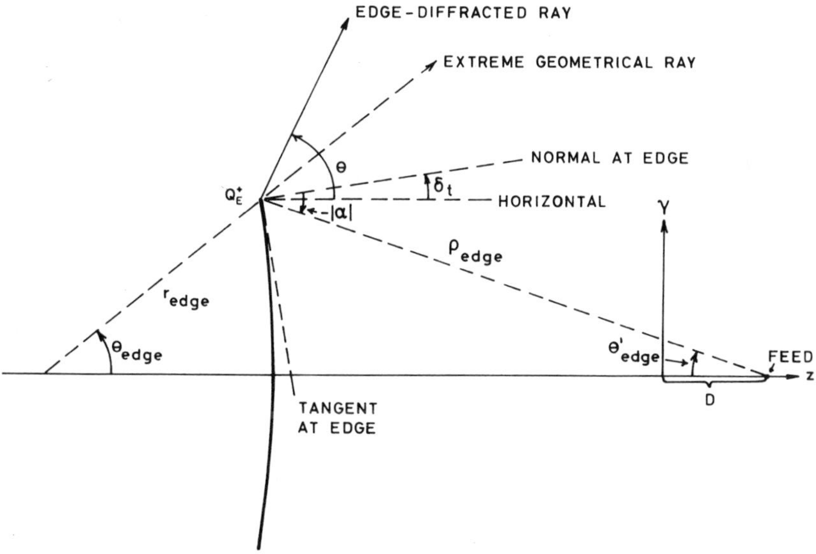

Fig. 7.10. Geometry of ray diffracted from Q_E^+

where [cf. (6.33)–(6.37)]

$$D^+_{s;h} = \left\{ -\frac{\exp(-j\pi/4)}{2\sqrt{2\pi k}} \cdot \left[\frac{F[kL^i a(\theta-\delta_t-\alpha)]}{\cos\left(\frac{\theta-\delta_t-\alpha}{2}\right)} \pm \frac{F[kL^r a(\pi+\theta-\delta_t+\alpha)]}{\sin\left(\frac{\theta-\delta_t+\alpha}{2}\right)} \right] \right\}, \tag{7.50}$$

$$a(x) \equiv 2\cos^2\left(\frac{x}{2}\right) \tag{7.51a}$$

$$F(kLa) = 2j\sqrt{kLa}\, e^{jkLa} \int_{\sqrt{kLa}}^{\infty} e^{-j\tau^2}\, d\tau\,. \tag{7.51b}$$

The transition function $F(kLa)$ removes the singularities at the shadow boundary ($\theta = \theta'_{\text{edge}}$) and at the reflection boundary ($\theta = \theta_{\text{edge}}$). Furthermore [cf. (6.46)]

$$L^i = \varrho_{\text{edge}}\,, \tag{7.52a}$$

$$L^r = \frac{\beta_1^r \beta_2^r \sin\theta_{\text{edge}}}{(D_{\text{refl}}/2)}\,, \tag{7.52b}$$

where β_1^r and β_2^r are the principal radii of curvature of the reflected geometrical ray at Q_E^+, and $(D_{\text{refl}}/2)\sin\theta_{\text{edge}}$ is the caustic distance evaluated in the direction of the reflection boundary.

For the point source on the z-axis illuminating the axially symmetric reflector, the resulting single-edge-diffracted rays have a caustic at all points on the Z-axis. Thus all ray path-lengths $\overline{FQ_EP}$ are the same if both F and P lie on the axis of symmetry. Ray optical solutions fail in the neighborhood of these caustics, as, for example, the $\sin\theta$ denominator in the square root factor of both (7.48) and (7.49) vanishes. Alternative solutions near caustics are canonical solutions [7.30], asymptotic solutions [7.31, 32], and integral solutions [7.33, 34]. The latter technique consists of replacing the edge-diffracted rays with contributions from equivalent electric and magnetic ring currents lying along the edge, specified, respectively, by [cf. (6.86) and (6.87)]

$$I_{e\phi} \cong -\frac{E_\phi D_s(\phi,\phi';\pi/2)}{\eta}\sqrt{\frac{8\pi}{k}}\exp(-j\pi/4)\,, \tag{7.53a}$$

$$I_{m\phi} \cong -H_\phi \eta D_h(\phi,\phi';\pi/2)\sqrt{\frac{8\pi}{k}}\exp(-j\pi/4)\,, \tag{7.53b}$$

where E_ϕ and H_ϕ are the edge-tangential incident electric and magnetic fields, $D_s(\phi,\phi';\pi/2)$ and $D_h(\phi,\phi';\pi/2)$ are the soft and hard edge-diffracted coefficients, and ϕ,ϕ' are the angles of diffraction and incidence in the "ray-fixed" coordinate systems defined by Kouyoumjian [7.29]. For an arbitrary field point ϕ is different at various points on the rim. However, for a field point on axis ϕ is constant (at great distances on axis $\phi=\pi/2-\delta_t$. Furthermore, for this symmetric case $\phi'=\pi/2+\alpha$.)

Axially Symmetric Reflector for a Vector Spherical-Wave Point Source off Axis

If a point-source feed is located off the axis of the axially symmetric reflector, Fermat's principle reveals that there are a maximum of four possible diffraction points around the rim of the reflector [7.35]. The situation becomes somewhat less complicated when the point-source feed, the reflector axis, and the observation point all lie in the same plane (taken, for example, to be the $X-Z$ plane in Fig. 7.11). Under these conditions, diffraction Points 1 and 2 are possible for all possible positions of the distant field point. However, if the field point is on the opposite side of the axis from the source point, then two additional diffraction Points 3 and 4 are also possible in the range of θ values, $\theta_1 \leqq \theta \leqq \theta_2$,

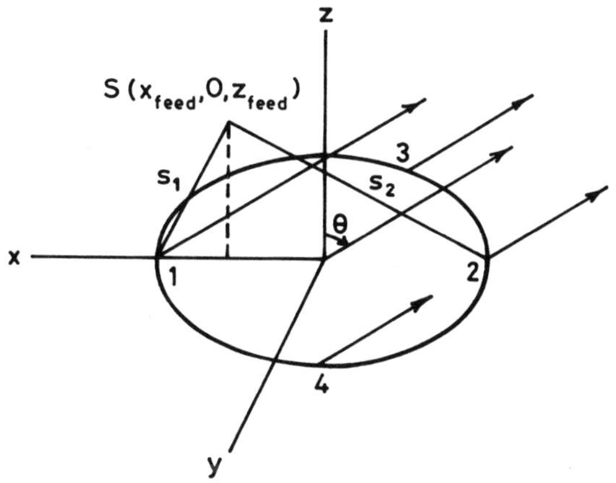

Fig. 7.11. Geometry of four possible edge rays

[7.36] where

$$\theta_1 = \sin^{-1}\left[\frac{X_{\text{feed}}}{\sqrt{(D_{\text{refl}}/2 + X_{\text{feed}})^2 + (Z_{\text{edge}} - Z_{\text{feed}})^2}}\right], \tag{7.54a}$$

$$\theta_2 = \sin^{-1}\left[\frac{X_{\text{feed}}}{\sqrt{(D_{\text{refl}}/2 - X_{\text{feed}})^2 + (Z_{\text{edge}} - Z_{\text{feed}})^2}}\right]; \tag{7.54b}$$

θ_1 corresponds to a caustic at infinity from Point 2, while θ_2 corresponds to a caustic at infinity from Point 1. Thus, as θ increases from θ_1 to θ_2, the additional roots 3 and 4 move from 2 to 1. In practice, for typical reflector geometries the range of values from θ_1 and θ_2 is sufficiently small that the entire range is covered by radiation from the equivalent ring sources to avoid unrealistically large predicted fields.

Surface-Diffracted Rays

First-order surface rays will ordinarily not be directly excited by the feed in a reflecting antenna system. However, they may be excited by tangential edge-diffracted rays. An example may be seen in Fig. 7.9, where an edge-diffracted ray from Q_E^- in the direction $\theta = \pi/2 - \delta_t$ will tangentially graze the surface. Only the tangential H-field component will be non-zero in this direction, but the excited surface rays are necessary to provide a continuous field. In most instances, however, the edge-diffracted rays provide a sufficiently accurate representation of the field without including the relatively more complicated surface rays.

Applications

The KOUYOUMJIAN group has used a combination of single edge-diffracted GTD, equivalent ring-currents, and physical optics to compute the complete radiation pattern of a prime focus paraboloid [7.33, 37]. The calculated pattern was compared with the experimental results of AFIFI, which were measured for a half-paraboloid on a groundplane with a monopole feed at the focus [7.41]. This experimental arrangement virtually eliminated aperture blocking. The vertical polarization eliminated the possibility of exciting surface-diffracted rays. The deep reflector ($F/D = 0.25$) eliminated rear spillover. The agreement between the calculated and measured patterns was very good between 0° and 130°.

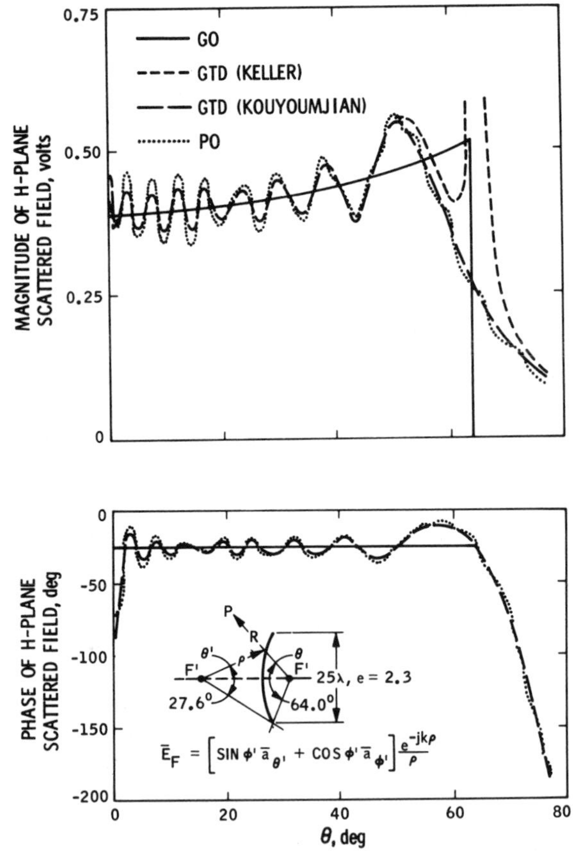

Fig. 7.12. Comparison of GO fields with other results for hyperboloid with feed at external focus

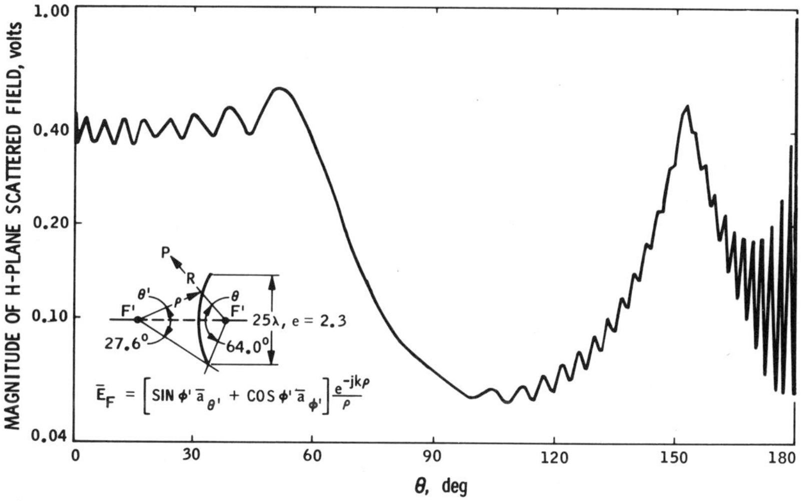

Fig. 7.13. Complete GTD H-plane pattern of hyperboloid with feed at external focus

Beyond 130° the agreement was also quite good, considering that at such low signal levels part of the discrepancy was due to measurement error.

The original Keller formulation of GTD [7.30] was applied to hyperboloids as early as 1967 [7.39]. The results compared favorably with the PO results, except for the singularities at shadow and reflection boundaries. The Kouyoumjian formulation with transition functions eliminated these singularities and yielded an accurate, rapidly computed radiation pattern. Typical amplitude and phase results [7.40] are plotted in Fig. 7.12 which compares GO, PO, GTD (Keller), and GTD (Kouyoumjian). The two GTD versions consist of the GO ray plus two singly edge-diffracted rays, with the Keller result exhibiting the singularity near the reflection boundary. In general, the PO and Kouyoumjian GTD curves agree closely in magnitude and phase, both in the illuminated and shadowed regions. The PO oscillations are somewhat larger than the GTD oscillations, indicative of the fact that two different edge current densities were assumed for the two results. In fact, in the H-plane the *E* vector is tangent to the edge at the point of edge-diffraction, and the actual current density becomes infinite at the edge for this polarization [7.41]. Of particular significance is the agreement between the PO and Kouyoumjian GTD results in the vicinity of the reflection boundary. Similar results are found for the E-plane.

Figure 7.13 is a complete H-plane pattern based on GTD. The fields at the reflection boundary at 64° and the shadow boundary at 152.4° are finite and continuous, and the two axial caustics are computed using the equivalent ring sources. This technique yields slightly different

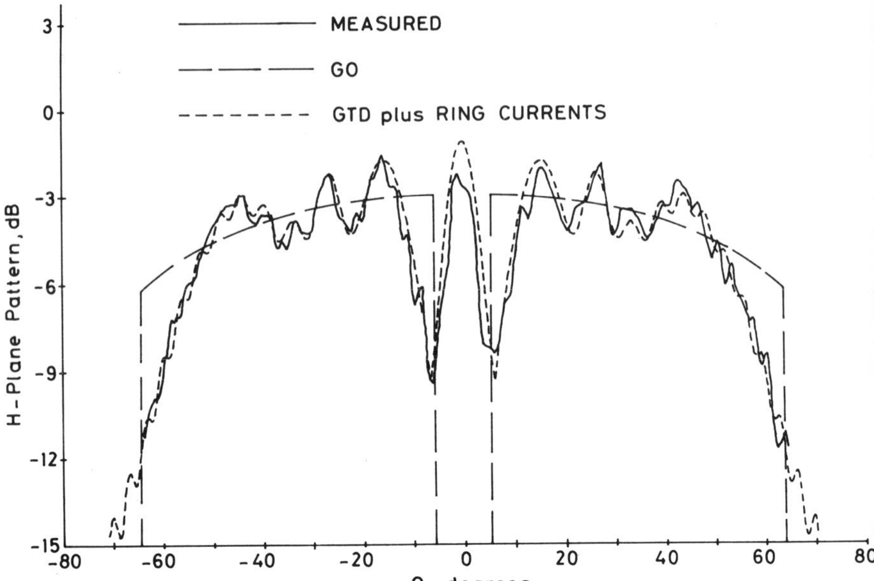

Fig. 7.14. Comparison of GO fields with measured results for hyperboloid with feed at external focus

results at the rear axial caustic than the caustic correction factor of KELLER [7.30] which did not yield E-plane–H-plane continuous fields on axis.

Figure 7.14 shows similar results for an 11.3-wavelength hyperboloid symmetrically illuminated by a 10-GHz corrugated horn [7.42]. In this case the GO results (heavy solid line), which now include feedhorn blockage, are compared with both GTD and experimental results.

The techniques of GTD provide a computational procedure to obtain rf performance data that hitherto was excessively costly in computer time. A prominent example of this is the dual-reflector antenna. Hitherto, it was only possible to analyze dual-reflector systems by integrating over both reflectors. The subreflector in a system of this type generally creates a shaped, rather than a focused, beam, and consequently is readily amenable to GTD determination of the scattered field. This rapidly determined scattered field then provides the illumination function for a physical-optics integration over the large, focused primary mirror.

The equivalent paraboloid is said to provide an accurate technique for the calculation of performance characteristics for Cassegrain and Gregorian systems [7.43]. GTD provides an accurate and relatively economical technique to verify this assertion quantitatively. For example,

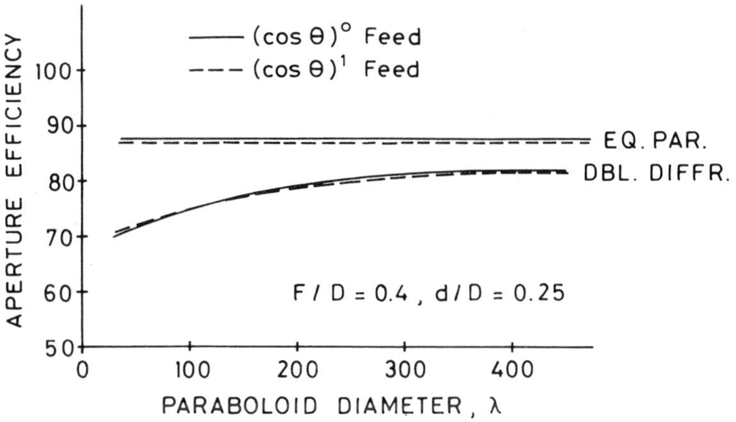

Fig. 7.15. Comparison of aperture efficiency values using GTD and equivalent paraboloid

in Fig. 7.15 the aperture efficiency (exclusive of spillover) is calculated for a Cassegrain system with a paraboloid $F/\mathrm{D}=0.4$ and subreflector diameter/main reflector diameter ratio $=0.25$. Diffraction from the subreflector causes the illumination of the main reflector to be tapered. However, this taper is not included in the equivalent paraboloid calculation, and its predicted efficiency is significantly higher, even for subdish diameters as large as 100 wavelengths. Since the edge illumination of the equivalent paraboloid is higher than the actual edge illumination, it generally predicts higher wide sidelobes than actually exist [7.44].

7.4. Application of Spherical-Wave Theory to Reflector Feed-System Design

It may be shown that an electromagnetic field $\boldsymbol{E}(\varrho, \theta, \phi)$ *and* $\boldsymbol{H}(\varrho, \theta, \phi)$ in a region outside a sphere (which encloses all sources) of radius $\varrho_0 > 0$ may be expressed as a superposition of vector spherical waves.

$$\boldsymbol{E}(\varrho, \theta, \phi) = -\sum_m \sum_n a_{e,o,m,n}\, \boldsymbol{m}_{e,o,m,n} + b_{e,o,m,n}\, \boldsymbol{n}_{e,o,m,n}\,, \tag{7.55a}$$

$$\boldsymbol{H}(\varrho, \theta, \phi) = (k/j\omega\mu) \sum_m \sum_n a_{e,o,m,n}\, \boldsymbol{n}_{o,e,m,n} + b_{e,o,m,n}\, \boldsymbol{m}_{o,e,m,n}\,, \tag{7.55b}$$

where the well-known spherical waves $\boldsymbol{m}_{o,e,m,n}$ and $\boldsymbol{n}_{o,e,m,n}$ were defined by LUDWIG [7.34]. If the tangential $\boldsymbol{E}(\varrho_1, \theta, \phi)$ is known on a sphere of radius ϱ_1 (where $0 \leqq \varrho_0 \leqq \varrho_1 \leqq \infty$), then the coefficients $a_{e,o,m,n}$ and $b_{e,o,m,n}$ can be uniquely determined. For example, determining the tangential components E_θ and E_ϕ at great distances from the sources will permit the complete field to be known for all values of ϱ subject to $\varrho_0 \leqq \varrho \leqq \varrho_1$. Consequently, this technique provides a unique analytical tool for transforming far-field data into near-field data, and visa versa.

7.4.1. NASA/JPL 64-m Antenna Dichroic Feed System

In support of the Viking Mars Project in 1967, and for science and technology demonstrations during the Mariner/Mercury mission in 1974, a dual-frequency (S- and X-band) feed system has been developed for the NASA/JPL 64-m antenna in California [7.46]. This system is capable of simultaneous low-noise reception at S- and X-bands and high-power transmission at S-band. To fulfill this requirement, a particularly interesting approach, the reflex feed system, has been implemented. A cross-sectional view of the reflex feed-system geometry is shown to scale in Fig. 7.16. The system consists of: the S-band feedhorn, an ellipsoidal reflector, a planar dichroic reflector, and the X-band feedhorn. By reciprocity, the operation of the reflex feed is the same in the receiving mode as in the transmitting mode; for simplicity, Fig. 7.16 shows only the transmitting mode. For S-band operation, from a geometrical-optics standpoint, radiated energy from one of the ellipsoid foci f_1

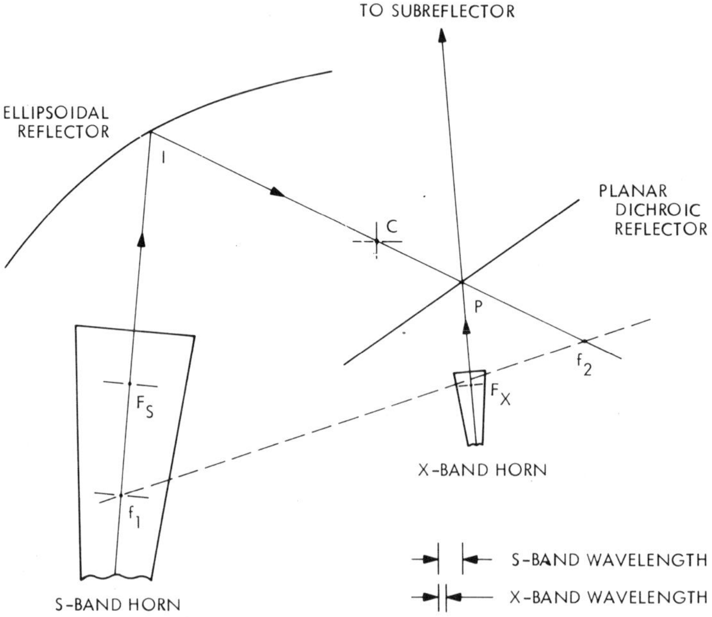

Fig. 7.16. Geometry of reflex feed system

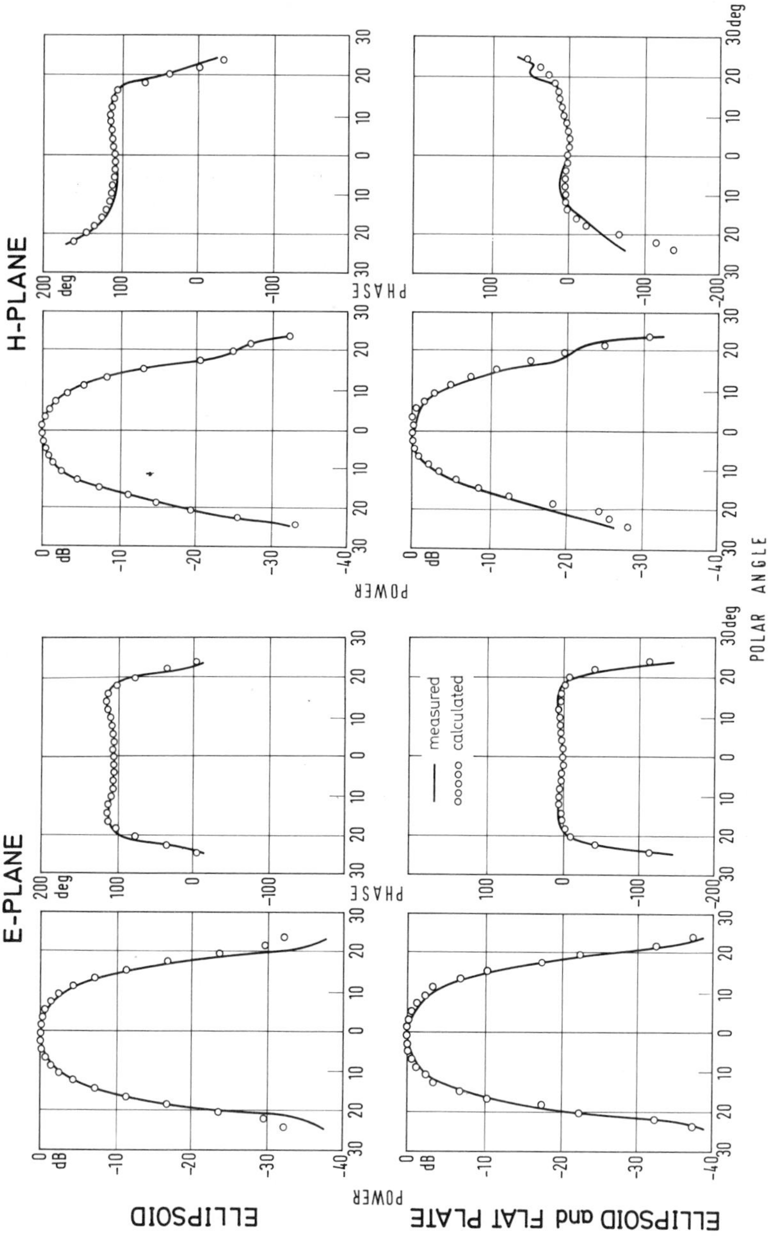

Fig. 7.17. Computed and measured patterns of reflex feed system

is focused to the Point f_2. As shown in Fig. 7.16, however, the system is not large compared to a wavelength. Because of this consideration and the fact that the S-band feedhorn does not represent a point source, the radiated energy from the ellipsoid is actually found to focus to a small region centered at the Point C. This energy is then re-directed by the planar reflector to the antenna subreflector. By the principle of images, this redirected radiation appears to emanate from the Point F_x, which is the far-field phase center of the X-band feedhorn and also coincides with one of the subreflector foci. To permit simultaneous X-band operation, the central region of the planar reflector is perforated with an array of X-band slots, thereby making the reflector essentially transparent to X-band but reflective to S-band.

Design of the reflex feed system was carried out with simultaneous 1 : 7 scale model testing and physical-optics scattering calculations based on spherical-wave transformations [7.47]. Sample calculated and measured field patterns are compared in Fig. 7.17. Because each

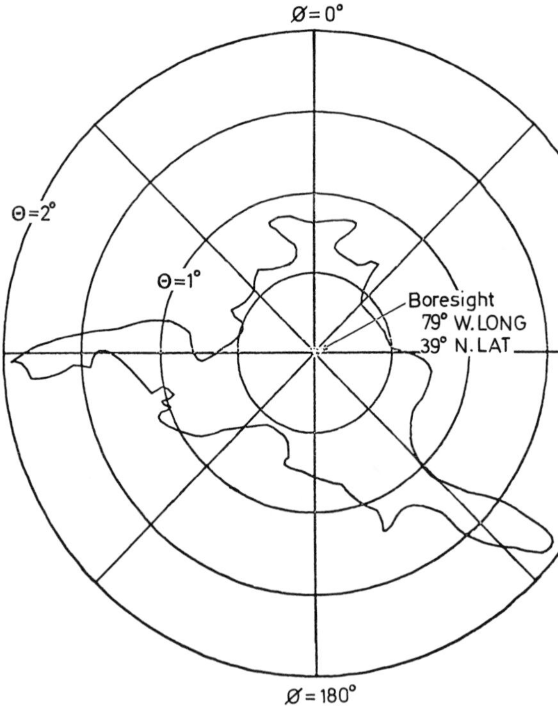

Fig. 7.18. Eastern U.S. time zone seen from synchronous equatorial satellite at 112° W longitude

reflection is essentially in the near-zone of the previous horn or scatterer, it is necessary to use far-field – near-field transformations. For example:

a) The far-field radiation pattern of the S-band horn was determined experimentally.

b) This field was transformed into its proper value (including radial field components) at the surface of the ellipsoid.

c) Using geometrical optics for the currents on the front of the ellipsoid, the far-field scattered amplitude and phase patterns have been determined and are plotted in Fig. 7.17.

d) This field is then transformed from the far-field to the vicinity of the surface of the planar reflector.

e) Geometrical optics then yields the currents on the planar reflector, from which can be determined the final scattered field, also plotted in Fig. 7.17.

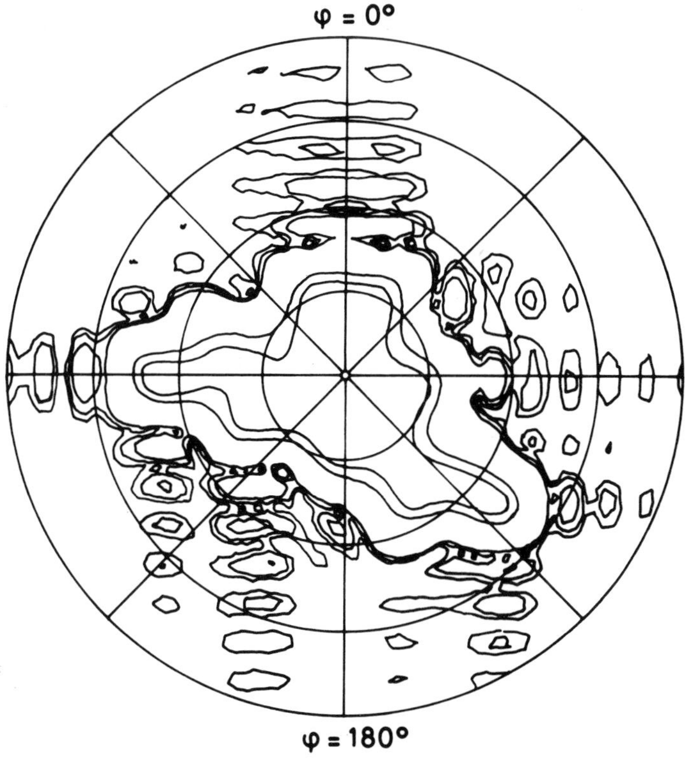

Fig. 7.19. Synthesized pattern using 16 odd and 16 even azimuthal modes

7.4.2. Shaped-Beam Pattern Synthesis Using Spherical-Wave Theory

WILLIAMS [7.48] has used spherical-wave transformations to synthesize complex pattern shapes, as example of which is shown in Fig. 7.18, the eastern time zone of the U.S. as viewed from a synchronous satellite over the equator at 112° W longitude (approximately due south of California). This geographic contour was then matched by a set of orthogonal beams from a finite aperture with an area of $160\,\lambda^2$. This physically realizable intermediate pattern was expanded in a series of 16 odd and 16 even spherical-wave azimuthal modes, yielding the pattern shown in Fig. 7.19. In principle, the technique described in Subsection 7.3.1 can then be used to synthesize an appropriate feed pattern.

7.5. Integral-Equation Analysis of Large Reflectors

The application of integral-equation techniques to the scattering from large reflectors has, as of this writing, been limited to two-dimensional reflectors [7.49, 50]. The E-wave solution from [7.49] is summarized in this section. The geometry has a plane of symmetry, so that only half of the reflector is shown in Fig. 7.20. The front of the reflector consists of 90 linear segments, each about 0.175 wavelength long, the endpoints

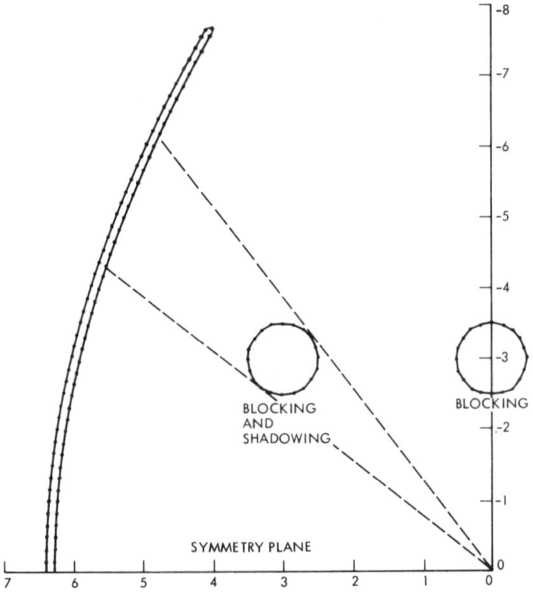

Fig. 7.20. Geometry of blocked and shadowed cylindrical parabola reflector

of which lie on a parabolic cylinder defined by the polar equation

$$\varrho_0 = \frac{6.27\lambda}{\cos^2(\phi_0/2)}. \tag{7.56}$$

Consequently, the front of the reflector approximates a parabolic cylinder with a focal length of 6.27 wavelength. The geometrical aperture diameter is 15 wavelengths. The reflector is 0.1 wavelength thick, and the back is subdivided into 90 similar segments. The edges of the reflector are semi-circular, of radius 0.05 wavelength, and add an additional 0.148 wavelength to the geometrical diameter of the aperture. Each semi-circular edge is divided into five segments 0.031 wavelength long.

The incident field illuminating the reflector consists of a line source located at the origin (which is also the focus of the parabola). The incident field intensity is given by

$$E_z^{\mathrm{i}} = \frac{\sec\left(\dfrac{\phi_0}{2}\right)\exp(-jk\varrho_0)}{\sqrt{\varrho_0/\lambda}} \tag{7.57}$$

in the direction of the reflector and zero in other directions. Consequently, the field incident on the front of the reflector is uniform in magnitude, producing (in geometrical terms) a uniformly illuminated aperture.

Calculations were carried out for the parabola alone, and the parabola in the presence of two symmetrically placed one-wavelength-diameter circular cylinders. In one case (blocking) the cylinders are placed in the focal plane so as not to intercept direct radiation from the line source but to affect (block) the scattered fields due to the currents induced on the parabola. In the other case (blocking and shadowing) the cylinders are placed closer to the reflector so as to intercept direct radiation from the feed before it illuminates portions of the reflector (shown with dotted lines in the figure) while at the same time blocking scattered radiation from the parabola.

The field of (7.57) was used to illuminate the reflector in the absence of the blocking cylinders. The current density induced on each reflector segment was computed and plotted in Fig. 7.21. Only the current density induced on the symmetric half of the reflector shown in Fig. 7.20 is plotted. The abscissa is the linear path length (in wavelengths) along the front and back of the reflector. The current density varies uniformly along the illuminated front of the reflector, has a moderate peak at the edge (light-shadow boundary), and drops by two orders of magnitude to a rapidly damped oscillation on the back. For comparison, the

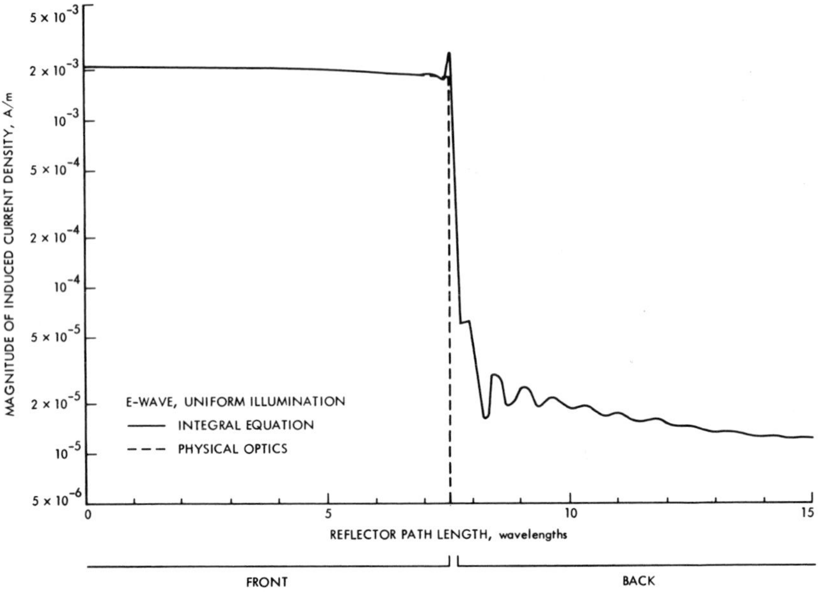

Fig. 7.21. Current density induced on unblocked reflector by feed with parallel polarization

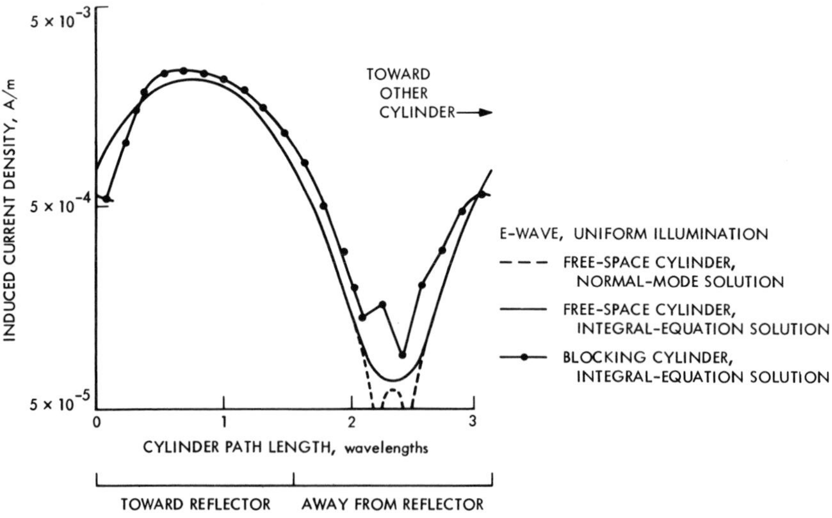

Fig. 7.22. Comparison of current densities induced on blocking cylinder by feed with parallel polarization

geometrical-optics approximation to the induced current density is also shown on the figure. The two currents are virtually the same on the illuminated front, except for a slight perturbation in the integral-equation result near the edge of the reflector. The physical-optics current drops to zero at the edge and remain zero on the back of the reflector. The current density on the shadowed back of the reflector rapidly decays 40 dB below its value on the front. It is not known whether the damped oscillations are computational or physical in nature.

Placing the pair of cylinders in the blocking position in the focal plane causes a modification of the currents on the parabola. The current variation on the front of the reflector changes from monotonic to oscillatory. Similarly, the oscillatory pattern on the back is significantly modified, although the magnitude is still approximately 40 dB below the currents on the front.

The density of the current induced on one of the blocking cylinders is plotted in Fig. 7.22. Near-zone fields of the reflector may be approximated using the geometrical interpretation that a plane wave is initially scattered

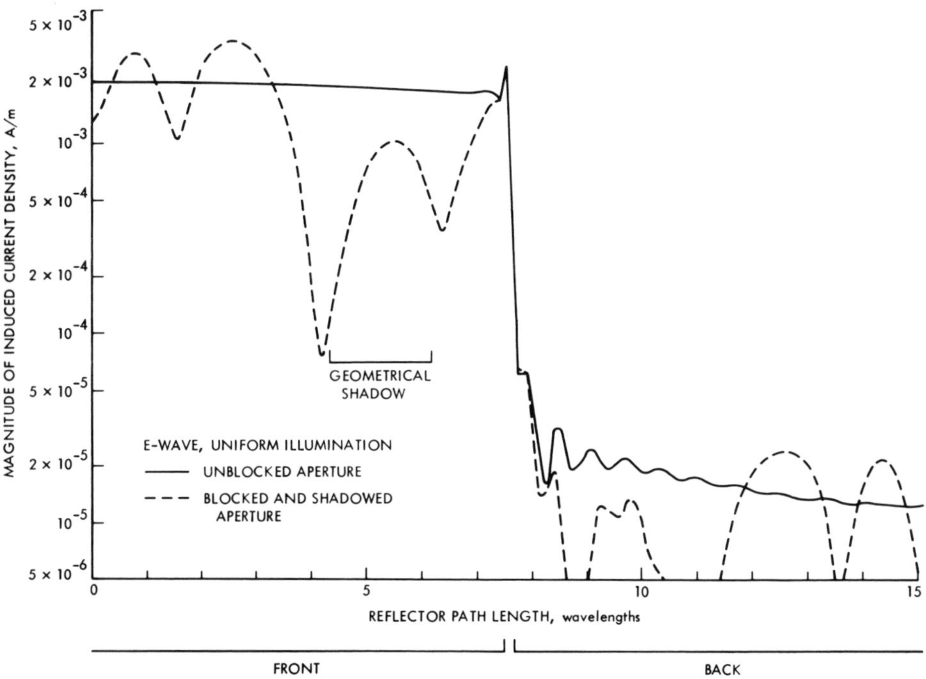

Fig. 7.23. Current density induced on blocked and shadowed reflector by feed with parallel polarization

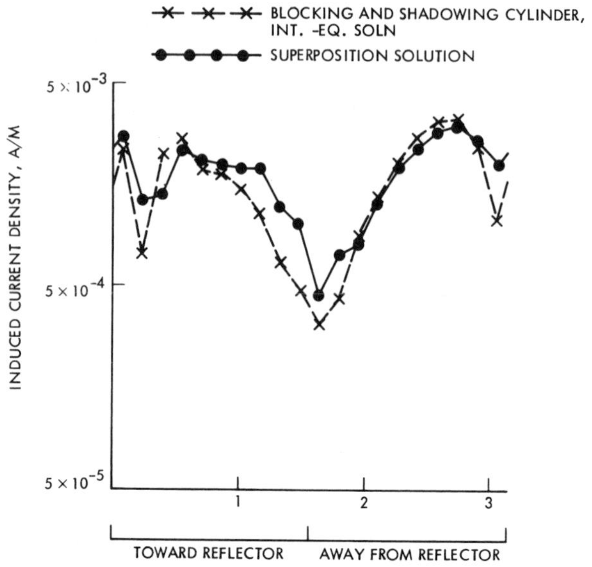

Fig. 7.24. Comparison of current densities induced on blocking and shadowing cylinder by feed with parallel polarization

from the reflector. Consequently, to provide a reference, the current induced on a free-space right-circular cylinder by an incident planar E-wave of a magnitude given by the plane wave scattered geometrically from the parabola is also plotted in Fig. 7.22 as computed both by the classical normal-mode solution and as computed by the integral-equation method. Except for low magnitudes, these two are indistinguishable. It may be seen that the currents induced on the blocking cylinders closely resemble those induced on a cylinder in free space by a plane wave of comparable magnitude, although the amplitude of the currents on the blocking cylinders averages 10–20% higher.

Finally, the problem was again solved with the two cylinders shifted forward so as to intercept radiation from the feed. The current induced on the parabola is plotted in Fig. 7.23, in addition to the reference current of the unblocked parabola. Strong oscillations are evident in both the illuminated and shadowed regions. Furthermore, the current in the shadowed regions of the reflector is significantly reduced although not eliminated.

The cylinder current distribution for this location of the cylinder is plotted in Fig. 7.24. The current was computed directly using the integral equation for the complete cylinder-parabola geometry. In addition, a current distribution was constructed by superimposing the current induced by the line feed in free space and the current induced by a plane wave equal in magnitude and phase to the quasi-planar wave reflected from the reflector. The superposition solution is a relatively good approximation to the integral-equation current density.

References

7.1. W. V. T. Rusch, P. D. Potter: *Analysis of Reflector Antennas* (Academic Press, New York, 1970), Section 2.6.

7.2. C.-T. Tai: *Dyadic Green's Functions in Electromagnetic Theory* (Intext Educational Publishers, Scranton, 1971).

7.3. W. V. T. Rusch, P. D. Potter: ibid., pp. 46–49.

7.4. P. Beckmann: *The Depolarization of Electromagnetic Waves* (Golem Press, Boulder, Colo., 1968), Chapter 3.

7.5. J. J. Bowman, T. B. A. Senior, P. L. E. Uslenghi: *Electromagnetic and Acoustic Scattering by Simple Shapes* (North Holland Publishing Co., Amsterdam, 1969), pp. 29–31.

7.6. S. Silver: *Microwave Antenna Theory and Design* (McGraw-Hill, New York, 1949), pp. 161–162.

7.7. W. V. T. Rusch, P. D. Potter: ibid., pp. 91–92.

7.8. J. W. Cooley, J. W. Tukey: Math. Comp. **22**, 297 (1965).

7.9. W. V. T. Rusch, A. C. Ludwig: IEEE Trans. Antennas Propagation AP-**21**, 141 (1973).

7.10. W. F. Williams: FY '71 Final Report, Communication Satellite Antenna Research, NASA Work Unit 164–21–54, JPL 362–10601–0–3330, Jet Propulsion Laboratory, Pasadena, Calif.

7.11. W. Romberg: Kon. korsk. Videnskab. Forhandl. **28**, 30 (1955).

7.12. A. C. Ludwig: IEEE Trans. Antennas Propagation AP-**16**, 767 (1968).

7.13. J. Focke: Ber. Sächs. Ges. (Akad.) Wiss. **101**, No. 3 (1954).

7.14. C. C. Allen: "Final report on the study of gain-to-noise-temperature improvement for cassegrain antennas"; General Electric Co., Schenectady, New York, 1967.

7.15. A. C. Ludwig: Private communication.

7.16. R. E. Collin, F. J. Zucker: *Antenna Theory* (McGraw-Hill, New York, 1969), Pt. 2, pp. 17–35.

7.17. F. S. Holt: "Application of geometrical optics to the design and analysis of microwave antennas"; AFCRL-67-0501, AFCRL, Bedford, Mass., Sept. 1967.

7.18. "Calculations of the caustic surface of a paraboloid of revolution for an incoming plane wave of twenty degrees incidence"; Parke Math. Lab., Concord, Mass., Rep. 1, Contract AF19(604)-263 for AFCRL, May 1952.

7.19. "Calculation of the caustic (focal) surface when the reflecting surface is a paraboloid of revolution and the incoming rays are parallel"; Parke Math. Lab., Concord, Mass., Rep. 3, Contract AF19(122)-484 for AFCRL, May 1952.

7.20. W. V. T. Rusch, A. C. Ludwig: IEEE Intl. Symp. on Antennas Propagation. Digest, Los Angeles, Calif. (1971).

7.21. O. Hachenberg, B. H. Grahl, R. Wielebinski: Proc. IEEE **61**, 1288 (1973).

7.22. A. C. Lunn: Ap. J. **27**, 280 (1908).

7.23. R. T. Jones: J. Op. Soc. Am. **44**, 630 (1954).

7.24. P. W. Hannan: IEEE Trans. Antennas Propagation AP-**9**, 140 (1961).

7.25. V. Galindo: IEEE Trans. Antennas Propagation AP-**12**, 403 (1964.

7.26. W. F. Williams: Microwave J. **8**, 79 (1965).

7.27. R. Lindsey: Aerosp. Techn. **17**, 28 (1965).

7.28. T. KITSUREGAWA, M. MIZUSAWA: IEEE Intl. Symp. on Antennas Propagation. (Digest, Boston, Mass., 1968).
7.29. See also R. G. KOUYOUMJIAN, P. H. PATHAK: Proc. IEEE **62**, 1448 (1974).
7.30. J. B. KELLER: J. Appl. Phys. **28**, 426 (1957).
7.31. I. KAY, J. B. KELLER: J. Appl. Phys. **25**, 876 (1954).
7.32. D. LUDWIG: Comm. Pure Appl. Math. **19**, 215 (1966).
7.33. P. A. J. RATNASIRI, R. G. KOUYOUMJIAN, P. H. PATHAK: Rep. 2183-1, Electro Science Laboratory, The Ohio State University, Columbus, Ohio (1970).
7.34. G. L. JAMES, V. KERDEMELIDIS: IEEE Trans. Antennas Propagation AP-**21**, 19 (1973).
7.35. F. MOLINET, L. SALTIEL: Laboratoire Central de Telecommunications, ESTEC Contract No. 1820/72HP, July (1973).
7.36. H. BACH, K. PONTOPPIDAN, L. SOLYMAR: Laboratory of Electromagnetic Theory, Technical University of Denmark, ESTEC Contract No. 1821/72HP, December (1973).
7.37. R. G. KOUYOUMJIAN, P. A. J. RATNASIRI: International Electr. Conf. Proc., Toronto, Canada, 152 (1969).
7.38. M. S. AFIFI: *Electromagnetic Wave Theory,* Pt. 2, ed. by J. BROWN (Pergamon Press, New York 1967), pp. 669–687.
7.39. W. V. T. RUSCH: JPL TR 32-1113, Jet Propulsion Laboratory, Pasadena, Calif. (1967).
7.40. W. V. T. RUSCH: JPL Interoffice Memo 3335-72-089, Section 333, Jet Propulsion Laboratory, Pasadena, Calif. (21. August 1972).
7.41. R. F. HARRINGTON: *Time Harmonic Electromagnetic Fields* (McGraw-Hill, New York, 1961), pp. 261–262.
7.42. O. Sørenson: Thesis, Laboratory of Electromagnetic Theory, Technical University of Denmark (1973).
7.43. W. C. WONG: IEEE Trans. Antennas Propagation AP-**21**, 335 (1973).
7.44. R. BOOTH: Private communication.
7.45. A. C. LUDWIG: IEEE Trans. Antennas Propagation AP-**19**, 214 (1971).
7.46. D. A. BATHKER: Agard Symposium Digest, Munich, F. R. Germany, 29-1 (1973).
7.47. P. D. POTTER: JPL TR 32-1526, **VIII**, Jet Propulsion Laboratory, Pasadena, Calif. (1972).
7.48. W. F. WILLIAMS: FY '73 Final Report, Communication Satellite Antenna Research, NASA Work Unit 164-21-54, JPL 362-10601-0-3330, Jet Propulsion Laboratory, Pasadena, Calif.
7.49. W. V. T. RUSCH: JPL TM 33-478, Jet Propulsion Laboratory, Pasadena, Calif. (1971).
7.50. J. A. KINZEL: IEEE Trans. Antennas Propagation AP-**22**, 116 (1974).

Subject Index